T0276180

CAMBRIDGE LIBRARY COLLECTION

Books of enduring scholarly value

Botany and Horticulture

Until the nineteenth century, the investigation of natural phenomena, plants and animals was considered either the preserve of elite scholars or a pastime for the leisured upper classes. As increasing academic rigour and systematisation was brought to the study of 'natural history', its subdisciplines were adopted into university curricula, and learned societies (such as the Royal Horticultural Society, founded in 1804) were established to support research in these areas. A related development was strong enthusiasm for exotic garden plants, which resulted in plant collecting expeditions to every corner of the globe, sometimes with tragic consequences. This series includes accounts of some of those expeditions, detailed reference works on the flora of different regions, and practical advice for amateur and professional gardeners.

My Garden in the Wilderness

This story of an Indian garden was published in 1915. Its author, Kathleen L. Murray, was living in the remote north-eastern region of Bihar in the home of her brother, an indigo producer, and some of her musings on life and gardening in India had already been published in the periodical *The Statesman*. She viewed this work as not as a guide, but 'merely a rambling record of some years in a garden' which combined European plants such as roses and sweet peas with natives such as cannas and beaumontias. Along with her gardening successes and failures over three years, the book provides insights into the life of the European woman in India - with no employment, and required to be both idle and aloof from the lives of the wider population. Murray's descriptive powers and enthusiasm for her garden make this book both enjoyable and evocative of imperial India.

My Garden in the Wilderness

Kathleen L. Murray

CAMBRIDGE
UNIVERSITY PRESS

University Printing House, Cambridge, CB2 8BS, United Kingdom

Cambridge University Press is part of the University of Cambridge.
It furthers the University's mission by disseminating knowledge in the pursuit of education, learning and research at the highest international levels of excellence.

www.cambridge.org
Information on this title: www.cambridge.org/9781108076708

This edition first published 1915
This digitally printed version 2016

ISBN 978-1-108-07670-8 Paperback

MY GARDEN IN THE WILDERNESS

MY GARDEN IN THE WILDERNESS

BY

KATHLEEN L. MURRAY

WITH ILLUSTRATIONS BY

SHEILA RADICE

LONDON

W. THACKER & CO. 2 CREED LANE E.C.

CALCUTTA & SIMLA THACKER SPINK & CO.

TO

THE BOY

WHO BELIEVED IN FAIRIES

INTRODUCTION

"MAN," said a writer in the *Spectator*, "was originally a pastoral creature; at the beginning of things he lived in a garden; a garden set all about with the wilderness into which he was soon to be driven, keeping always, through his exile, some memory of his first estate, so that in the middle of the wilderness he will still make himself a garden."

I did not know that I was obeying a primitive instinct in making myself a garden in my wilderness; I imagined that love of beauty and of order were the chief incentives to work that was often disappointing. But now the thought comes that the instinct goes deeper, and that we are all, mentally, creating for ourselves some little, lovely garden in our own especial wilderness—gardens of dreams, of hopes, of ambitions, of love—and that without these gardens in the wilderness life would scarcely be tolerable, and exile be bitter indeed.

I would ask that this little volume be not accepted as a guide, even by beginners. It is merely a rambling record of some years in a garden or gardens, for there are two gardens about which my memory clings, and the thought of which will always be pleasant to me, although both are now gone from me for ever.

K. L. M.

FAIRY TALE TREES.

CHAPTER I

I know not how it is with you
I love the first and last—
The whole field of the present view,
The whole flow of the past.

Our lives and every day and hour
One symphony appear,
One road, one garden, every flower,
And every bramble dear.

R. L. S.

THE chandan* trees have just finished flowering and we breathe freely again. I am not sure that I should have chosen them for my garden, but long ago some one planted a group outside our gates and we allow them to remain, in spite of the disadvantage they entail in March and in October, when for two days the house and garden are

* The Indian sandalwood.

permeated and rendered almost uninhabitable by the pungent scent of their blossoms. But it is a cleanly odour, with an aromatic suggestion of health, almost like that of pine trees, or eucalyptus, and they are the quaintest trees I have ever seen, regular fairy tale trees with gnarled trunks and straight branches that grow stiffly upwards, and covered with precise bunches of dark green leaves. They suggest a decorative frieze by Walter Crane, and when I stand under them and look up into their branches, I seem to have left the common workaday world far behind me. And yet it is all around me, for the group marks the point at which the road branches, and, rounding the tank at the end of my garden, disappears from my view. I am glad to lose sight of it, for it is not a pleasant road, being deep in dust in summer and in mud in the rains, and down it come creaking carts, and harassed bullocks, and overladen pack ponies, that give me a pain at my heart.

In my garden there is a green refuge from the dust and glare outside. Behind the fern house the ground rises to the bank, and there, shaded by teak trees and closed in by a shrubbery, I have placed a seat. Even on hot mornings it is cool here, and I bring the dogs and a pretence of needlework and sit watching the casuarina trees

A GREEN REFUGE FROM THE DUST AND GLARE OUTSIDE.

balancing their delicate spires against the pale turquoise sky. We have groups of casuarinas planted at each end of the bungalow, and the least wind causes them to sway and sigh with a sound as of waves on a pebbly shore. The dogs are distinctly bored by my tendency to loiter. They long to be scampering down the dusty road and barking up into the twisted sissoo trees that border it, returning in kind the virulent abuse heaped on them by squirrels who sit in high places. But as dogs are the quintessence of politeness, they lie about with the air of attentive resignation that they know so well how to assume, and, I suppose, hope that their devotion will, in time, touch me into doing my obvious duty to them.

My shrubbery is a mixed one, for I have not yet learnt the true landscape gardener's art of massing colours. So hibiscus, scarlet and salmon and cream, jostles *Tecoma stans* with its lime-green foliage and bunches of yellow bells, and on the fern house roof a mass of heavenly purple contends wildly with the tawny *Bignonia venusta*. I shall never be a real gardener because I love flowers so much, all of them individually, no matter where they are. Real gardeners harden their hearts, they do not permit pirate poppies and petunias to live in their rose beds, and no

MY CLIMBING ROSES.

real gardener would for a moment dream of allowing the all-conquering morning glory to behave so badly to the more delicate, legitimate creepers on the fern house. I wish morning glory had not such fierce, unpleasant ways; I feel helpless before it. It creeps in through the latticework of the fern house and strangles the young begonias and maidenhairs, and it gets through the roof and worries the plants in the hanging baskets, and clings to my hair when I pass through. Some day I think I shall plant morning glory next to *Beaumontia grandiflora*, and then we shall see what happens. But perhaps beaumontia is too dignified to contend with a mere weed; there is something regal about the plant as there is about its name, a name I love to repeat to myself, it has so noble a sound. I have a sense of being honoured when, in February, the beaumontia gives me her great white, faintly scented blossoms for the yellow Doultonware bowls in my drawing-room. Perhaps it is because she flowers only then, and not, as many shrubs in India appear to do, whenever the fancy takes them. We have a coquettish kind of jessamine tree that is always taking us by surprise, and bursting into a wild revel of blossom on the excuse of a casual shower.

* * * * *

There are no graves in my garden, a fact worthy of remark, for few indigo factories are without these sad memorials of the Europeans who lived and died there. I suppose this place is not old enough to have them, for the next factory has a row of big masonry monuments down the side of the vegetable garden. The oldest of these graves is dated 1812

BEAUMONTIA.

on the slate slab let into the side, and is that of a lad, aged twenty, whose parents lived at Berhampore. That is all that is now known of him, for his name is not one of those on the factory record, which is in some sort a record of my own family for the last two generations.

Away from that garden, across a stretch of rough grass, and almost hidden amidst clumps of bamboos, lies a small walled graveyard with three nameless graves. From the roughly moulded urn on the flat masonry slabs we know they are Christian graves, but we can only surmise that they are those of Dutch pioneers, for the old factories were in the hands of the Dutch long before English people came to Behar. They refined saltpetre then and had nothing to do with indigo, which started here some time about 1780. I wonder sometimes what the place was like in those days, and how and where they lived, those rough Dutchmen. Was it in the long rambling house I know, to which rooms seem to have been added haphazard, for that is very old, too? Or was even the oldest portion of that built long after they had been laid to rest in the little quiet enclosure where the grass now grows rank about the cracked grave stones, and a big peepul tree has root in the crumbling wall? My imagination works vividly around this spot where

> "Only the sun and the rain come
> All year long."

* * * * *

But, having no graves nearer than six miles, I am this morning exhausting my speculative faculties in wondering what green parrots want with hibiscus flowers. A flock of green parrots has haunted my garden for some days past, and this morning they settled on a bush of scarlet bell hibiscus, and when they left it most of them had a great flower in their beaks. They flew into the tallest casuarina tree and perched there, apparently playing some delightful game, till, on a sudden alarm, they all flew chattering and screaming away, dropping the flowers on the path, where they now look like giant spots of blood. I picked one up, and it had only two little dents, one on each side of the base of the flower, so they never meant to mangle or to eat them. It must be delightful to be an emerald hued creature, able to sway on tree tops, with a scarlet flower in your mouth, conscious of your decorative value against a blue sky. For I don't for a moment believe that those birds did it for anything but effect. I have a green parrot of my own, and I know their subtle ways.

My parrot is not a happy bird, nor am I happy in possessing him, for I do not like to have caged creatures. But I bought mine long ago, in hope of saving him from a worse fate—a poor little thirsty fledgling in a

MY ILL-TEMPERED GREEN PARROT.

cage the size of a mousetrap. Now I am afraid to let him go free, for fear his fellows should peck him to death. I take care to supply him with sugar cane, guavas, and all the seasonable fruits, but I feel there is not much sympathy between us. He is vicious and noisy, and his one hope in life is that I shall one day incautiously put a finger far enough into his cage to be seized and mangled.

THE HOUSE OF THE CASUARINAS

CHAPTER II

Look to the rose that blows about us, "Lo!
Laughing," she says, "into the world I blow,
At once the silken tassel of my purse
Tear, and its treasure on the garden throw."

OMAR KHAYYAM.

THE first cold nights—and they came early this year—painted the tips of the amaranthus in my garden with flaming scarlet. Early last month I planted a row of these along the edge of the shrubbery that encircles my retreat at the back of the fern house, and now they are a joy to the eye. For all that, I find it impossible to respect any such splendid thing that is in reality so lacking in backbone, each plant requiring the invisible support of a split bamboo to enable it to stand upright at all.

* * * * *

I wish my dogs did not so strongly disapprove of "pottering," because I am at present too busy over my roses to take them for the walks they love. So they sit about, and hang out their tongues, and look bored, which gets on my nerves. The latest addition to my pack, a black-and-tan dachshund, is the exception. Her short legs, which render walks a weariness and squirrel hunting a tantalising impossibility, serve her in good stead in her self-imposed task of helping the *mali* * to open the roots of the rose bushes, a process into which she flings herself with spirit and the amiability that is characteristic of her kind. There is at the present moment an air of most joyous helpfulness about as much of her as I can see; while from the hole in which she has half buried herself there issues at intervals a shower of earth, accompanied by a choking snort that I take to be an expression of happiness in the dachshund language.

* * * * *

She is an ingratiating little person, and I fear that I am rapidly degenerating into the species

* Native gardener.

of bore that any dog lover who has ever owned a dachshund is bound to become. The only circumstance that may serve to arrest me in my downward career is the knowledge that I scarcely enter into her mental horizon: her every thought

THE MALI. AN ERROR OF JUDGMENT ON THE PART OF THE PLANT.

and her whole heart being given to the man of the house, from whom she has never, so far as I can see, received any kind of encouragement.

But I started out to talk of roses, not of dachshunds, and that brings me to wondering why, in this part of the world where lies my garden, roses are so little considered. I know many beautiful Behar gardens where mammoth chrysanthemums are the pride of their growers' hearts, and pansies, sweet peas, and carnations make gay the garden beds all the cold weather. But in most of these gardens the roses are left to the ungentle care of the *mali*, who prunes them cruelly, waters them infrequently, and otherwise maltreats them. I have even seen them relegated to the kitchen garden. Yet roses repay a little trouble—and they ask so little! To begin with, they flower practically the whole year round, only going to rest during the intense heat of May and June. Directly the rains break they are again in flower, giving smaller, rosette-shaped blossoms, it is true, but sweet-scented and plentiful. It is a mistake to suppose that they should then be prevented from flowering; the bush has had its rest in the hot months, and I have never found that I had less perfect blossoms in the cold weather because I had not the heart to clip off the buds in the rains.

* * * * *

My roses are in the very middle of the garden, planted in broad borders each side of the path that leads to the fern house, with a clear stretch of lawn on either hand. For roses must have sun, morning and afternoon. They are bold flaunting flowers, and will not grow in shady nooks or inconspicuous corners; they demand the best place in your garden; given that and water every day, they require very little other attention until the pruning time comes round.

* * * * *

I have just finished pruning, by which I do not mean cutting off all the shoots to a uniform height of two feet from the ground, which the *mali* contends is the proper method. I have cut out the old wood, and such shoots as I take to be the stock asserting itself, and it is no easy task to determine what is stock and what the more delicate grafted plant. I have no doubt I have cut several shoots that I should not have done, but only experience brings the nice discrimination necessary to all garden work.

* * * * *

Directly the pruning was over the roots were opened, and very old manure from the cow-shed was put in; then they were closed at once, and to-morrow they will be watered. Again the *mali* disapproved. He wanted to leave the roots exposed to the air for ten days, because, he assured me, this is the *dustoor*.* It is a custom I do not like at all, and I feel sure that the rose bushes like it no better than I do. I can well imagine them nodding their stems to one another, and saying in

little shocked whispers, "They do these things better in France." For most of my plants came from a "roseraie" near Lyons, though they appear to like the climate of India very well, and flowered most obligingly the first year I put them down. They were piteous-looking little things when they arrived by post, forty in a torpedo-shaped parcel,

* Custom.

with their stems clipped to a foot long, and no earth on their poor bare roots. Acting under instructions from one who knew, I plunged them at once into a thick mud bath, and planted them in new soil in a shaded place. In a fortnight there were leaf buds on the dingy mud-cased stems and soon they were in a glory of leaves, with little tentative buds opening shyly into a new world. They were tiny flowers at first, but of beautiful delicate colours, and as the cold weather went on—for they arrived in November—the blossoms grew bigger and of better form. At the end of their first year I planted them out in the rose walk with my other roses, and now they are acclimatized, and seem on very good terms with their sisters from Lucknow.

* * * * *

The French roses are all hybrid teas and hybrid perpetuals, which are really the most satisfactory roses for India. Of these last one of the most vigorous and beautiful is a German lady, Frau Karl Druschki, who has long, thick waxen petals of pure white; very lovely in the bud, but, like many ladies of her nation, a little inclined to look blowsy when full blown. She, however, remains in bloom a long time, although in this respect she cannot equal Mildred Grant, an indefatigable young person, whom I have actually seen in bloom for a week without having dropped a petal. It is on this account, I presume, that the catalogue describes her as an exhibition rose.

* * * * *

Paul Lede is a hardy rose, with profuse flowers of a pale terra-cotta tint, and another quaint rose is Le Progres, of a soft pastel yellow, described as "Nankin," but approaching apricot to my mind. Amongst my last year's parcel of roses from France (which I have not yet seen in cold weather bloom) is one that particularly charms me—Grüss an Teplitz. The flowers are deep ruby, with closely set petals, and a scent like all the sweetest scents you have ever smelt. I feel sure that his cold weather blossoms will be glorious, and I am paying especial attention to his bedding out which is now going forward.

* * * * *

After writing and thinking of roses, it is a little difficult to consider any lesser flower, and especially zinnias. Roses are so tender and gracious, and there is nothing of either quality about zinnias, which are cold, precise, self-confident flowers. Yet I like zinnias, and cultivate them in great masses, for the sake of their glorious colours. This year they were extremely late in flowering, and at present are usurping the narrow beds on either side of the drive that are the usual right of the chrysanthemums. I had to find other places for these last, and now I am sure the zinnias are feeling triumphant; they are the kind of creatures who would show it, too. Encouraged by the heavy night dews, they show no signs of flagging yet, and evidently mean to hold their places for another fortnight at least.

* * * * *

And now I think I must stop writing and try to tidy up the lawn, where Don, the terrier who will never grow old, is busily killing a cob of Indian corn, while the dachshund has ceased her digging to gaze in sheer admiration and envy at one so recklessly courageous!

THE TERRIER WHO WILL NEVER GROW OLD.

THE RED COTTON TREE.

CHAPTER III

> There's night and day, brother, both sweet things; sun, moon, and stars, brother, all sweet things; there's likewise a wind on the heath. Life is very sweet, brother, who would wish to die.
>
> GEORGE BORROW.

I HAVE been reading 'Lavengro,' a quaint, delightful book that is restful reading in spite of its prodigality of adventure, and that seems to harmonise with my lonely garden as few other books could. Romance, like beauty, is for the "seeing eye," and Borrow did not need to go far afield for his; it met him at every turn. The Flaming Tinman, vengeful Mrs. Herne with her fatal "drows," Jasper Petulengro with his sad philosophy of death, would have appeared but troublesome vagabonds to the ordinary mind.

Borrow touched them with mystery, and mystery is at the heart of romance.

* * * * *

The cold weather has come. Standing at my window in the early morning I look on to a garden veiled in delicate silver mist, from out of which loom the gold masses of the chrysanthemums, the scarlet of the poinsettias. The air is crisp and chilly; the cups of the roses are heavy with dew and spiders are hanging spangled webs on the rustic arches that span the path to the fern house. There is a long bed of violets just below my window, and beyond that a row of

pots of carnations which are just breaking into flower—last year's plants, of course, treasured during the rains.

I have forgotten the hot weather already; for me zinnias never bloomed, punkahs never swung; skies never blazed in the crude glory of an August sunset. The dogs have forgotten, too, and tear madly round the garden, and roll on the dewy grass, and come in muddy and boisterous and full of the joy of living.

We have *chota hazree* * on the verandah, and thank heaven for the thickness of the cream and for the big bunch of fresh roses in the centre of the table. If it could be always December in Behar! For the weather here in November and December comes as near perfection as can be imagined—such sunny still days; such clear, cold nights. In January the cold winds begin; by February the winds grow warm and soon the cotton trees are in bloom, and the cold weather is over.

* * * * *

I have said there are no graves in my garden. For all that there are ghosts; shadows that come and go along the paths and across the lawn; shadows of friends who walked with me; of dogs I loved, and more than all the others, daily with me,

* Small breakfast.

the boy who believed in fairies. He is a public school boy now, a "Blood" of the Upper Fifth, babbling of prowess in the cricket field; but in my garden I see him always a sailor in miniature,

THE VERANDAH.

spreading feasts for fairies under the clumps of canna by the fern house. The curled, orange-coloured cup of the canna seemed to him a comfortable and appropriate home for the "little people"; beneath the broad leaves he laid offerings of sugar—of gaily coloured scraps of cloth to clothe the fairies; later, of letters in his best, straggling hand. Ants ate the sugar, the cloth and letters were gathered up, and one morning—cold and dewy like these—saw a small, scarlet-faced being hurling itself frantically across the lawn and shouting "Mother! fairies are true!" The fairies had replied—a few words of gratitude printed on such tiny paper as fairies might be expected to use.

I have never defended myself. It was easy to foster such sweet illusion; even though in doing so I was, so I was told, unwise to the point of wickedness. Childhood is sadly full of disillusionment, if we but knew. Think of the tragedy of the moment when some older child whispers, "Santa Claus is only your own mother really. Watch and see," and then childishly elaborate traps are set, and all is over. I once heard a small boy say rather sadly after his first term at school, "I used to think mother knew everything; now I find she does not know algebra!"

* * * * *

Jerry is another shadow that walks with me. Insignificant, bad tempered, faithful and affectionate Jerry had a place in our hearts that his half brother Don, still beside me in the flesh, will never usurp. Yet Don is a creature of manifold fascinations; a spurious bull terrier with the lazy, appealing manners and the indomitable courage of his kind. Brought up apart the brothers con-

THE CANNA IS A GREEDY CREATURE.

ceived for each other at first sight an antipathy that, at intervals, brought the whole establishment on the scene armed with jugs of cold water, pepper castors and other expedients usually resorted to in the case of two dogs locked in a furious struggle. As time went on and the fights gradually ceased, it was seen that Don, although physically the stronger, had given Jerry "best," a veritable triumph of mind over matter. Yet when later there occurred one of those tragedies known to all dog lovers, and a merciful gunshot had terminated Jerry's ravings, Don mourned and missed his conqueror, searching for the little wise, white figure that would accompany our walks no more.

* * * * *

"We make our friends, we make our enemies, but God made our next door neighbour," so we are assured by G. K. Chesterton. I am convinced that the whimsical seeker after paradox once lived in the Indian *mofussil.** In England you may, and generally do, ignore your next door neighbour. The wall that separates your bodies divides your destinies as well, and so long as he is not noisy, and does not make himself in any way objectionable, you accept him as you accept the trees in the square, and the cat on the roof. In India your neighbour, especially when he happens to be the only white man within five miles, becomes an

* Provinces.

enormous factor in your existence. You are drawn into a dreary and unprofitable intimacy with him; you drink his whisky and proffer him your mutton chop; you even nurse him in sickness and are called upon to rejoice with him in health. While your souls are really leagues apart, there grows up between you the familiarity of propinquity, and the real horror of the situation is that there is no getting away from it. The shifting, changing life of the up-country station is not for us of the district, and it is very probable that our neighbour is ours unalterably, and that we shall continue to share his mutton, book and polo clubs until such times as one of us makes a fortune or dies.

* * * * *

The foregoing remarks do not mean that I bear the slightest ill-feeling towards my own neighbour, who is honest and kindly and sporting, and whom I suspect of being infinitely more bored with my conversation than I am with his. In any case he will never know that I have written about him because he looks upon me as a domesticated person who would never do anything so unbecoming as to write for the papers.

* * * * *

I love chrysanthemums, but they make me feel homesick. When I look at them my sunlight garden fades away, and I see a grey London street, with frowsy flower-women hawking great masses of the brilliant blooms at the street corners. We are all homesick, more or less, in this "land of regret," but I think the "London sick" are most to be pitied. The contrast is too sharp. Think of an autumn afternoon in the West End! The dull, soft light in which all women look pretty, the brilliancy of the shops. The theatres give up their throng, the tea shops are full. "Like dragon-flies the hansoms hover, with jewelled eyes to catch the lover." For a moment I am in the midst of it all, and then . . . . it is gone and there is only the afternoon peace of my garden; no jingle of hansom or cry of newsboy, only the caw of the homing crows and the shadow of the casuarinas on the sun-dried lawn.

THE OLEANDER.

CHAPTER IV

Across the grass sweet airs are blown
Our way, this day in Spring.

D. G. ROSSETTI.

"OH, to be in England, now that April's there!" There are few exiles who have not been thrilled by that homesick cry across the sea, even though Browning wrote from the comparative nearness of sunny Italy:

"And after April, when May follows,
And the white throat builds, and all the swallows"—

Even the words, without the sense, are beautiful, fraught with music and with longing.

I do not know if any poet has sung the praises of the Indian spring; indeed in most parts of the plains there is no transitory stage. The cold weather ceases to be, the hot winds blow, and summer descends. But in Behar we boast of spring; a space before the west winds grow too riotous; a season of cool nights and warm days, when the garden beds are gay with English flowers, and the young sissoo trees deck themselves in delicate lettuce green.

For us no show of daffodils, no long, sad spring twilight, no scent of wet earth in the sprouting fields, no glory of blossoming orchards; instead, the flowering of the cotton trees, the heavy fragrance of mango blossom, the azalea-like blossoms of the bauhinia, and the suggestion of

promise that is inseparable from Spring in any quarter of the globe.

* * * * *

As I sit at my window this morning I could imagine myself in Japan. At the edge of the tank that bounds my garden a great straggling cotton tree, its blood-red flowers closely massed on leafless branches, sprawls against a background of turquoise sky. Before it stands a delicate bauhinia, also leafless, and a flower in pale mauvy pink. To the right a cactus, very stiff and prim, occupies an exalted position on the bank of the tank and, valueless by itself, serves to complete the fascinating effect. Even the dachshund disporting herself on the lawn, in passionate pursuit of a butterfly, although incongruous, is, at least, picturesque!

* * * * *

"In England—now!" They are selling branches of mimosa in the London streets; gold against the grey, and the smoke-blackened trees in the Park are putting forth their leaves. Even the display of spring millinery in the shops adds to the sense of gaiety and hopefulness that is in the air.

Mimosa, which is the Indian babul, the Australian wattle, has such a strange, haunting scent. There are some rough, twisted babul trees outside my compound, and when I pass them I am reminded of Kipling's Australian trooper in South Africa who

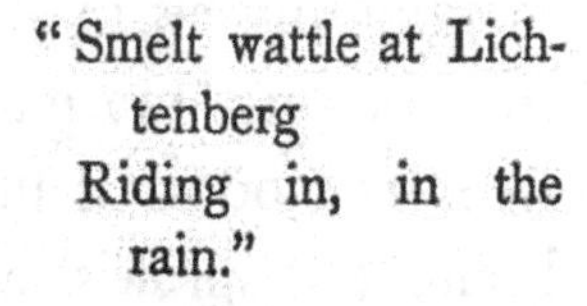

"Smelt wattle at Lichtenberg
Riding in, in the rain."

There is pathos in the rhythm of the words, and I know well the sudden overwhelming nostalgia that is aroused by a familiar scent.

* * *

I want to grow dwarf trees *à la Jap.* I have been seized with this ambition at intervals during my years in the wilderness, but relinquished it of necessity when told that a few decades go to the proper growth of these delightful Liliputians. Now some one tells me that there is a quite simple and comparatively speedy method, connected in some way with the planting of a seedling in an orange skin, and I want to hear more about it. To have a gnarled oak or even a hoary peepul on one's dinner table would be a distinct achievement. I don't hesitate to say that I should be properly proud of it, and should expect all my friends to assure me of the cleverness and originality of the notion.

By the by, I have never been able to understand why clever people are popularly expected to be oblivious of their own cleverness. It would argue a distinct lack of perception if they were, but I think one can safely assert that they never are oblivious. There is no harm in thinking oneself clever; the harm only comes in when we begin to show others how stupid they are by comparison.

* * * * *

Fashions in flowers are almost as evanescent as fashions in millinery. It is not long since canna was the rage; every garden I know was full of it, and people who had never gardened before became pronounced bores on the subject of canna, which appeals to the beginner as being a nice easy plant to grow. You planted it in any corner of your garden, the newer the soil the better, and no sooner had you planted it than—hey presto! it flowered gorgeously, sending up a quantity of suckers for the furtherance of its species. In fact, canna seemed almost too good to be true.

It was barely a year before a doubt crept in. The canna groups had a way of getting untidy, and the *mali* forgot to cut down the stems that had flowered. In time it dawned on us that canna was a greedy and ungrateful creature, an insatiable feeder, a thirsty drinker. It speedily exhausted the soil about it, and clamoured for more. We discovered that canna loved change, and must be transplanted every three months or so, and gradually we fell out of love with it. It was still gorgeous, but so easily propagated that every garden blazed with it; we wearied of its magnificence. And so canna fell into its proper place. It serves to fill a space, and it flowers often when there is little else in the garden, but it is in no sense to be relied on, and it is deceptive in that it requires more care than it professes to.

Cosmos is now the fashion, and its starlike blossoms are dainty and easy to arrange. But, like canna, it requires the very best soil in your garden, a rather special soil, to be seen to its full advantage.

* * * * *

All people with hobbies are bores, but the chief bores are gardeners and dog lovers. If you add to these tastes a love of books you become a bore of the first water. It is as well that I live in a wilderness, and, with few exceptions, the few visitors who penetrate to my fastnesses treat my tastes with tolerance, and the dogs with the deference they have learnt to expect.

COSMOS IS NOW THE FASHION.

We had a duty visitor last week, and I was sorry for her, although she depressed the dogs by alluding to them as "animals," an epithet from which they have hitherto been carefully guarded. She ignored the garden and complained of the noise of the birds, but she took me through the history of three love affairs—her own—and a discourse on the jealous nature of women. She added that if the story of her life were written it would make an interesting book. I have heard that remark before, and it is one that requires no comment. We parted with, I am convinced, mutual relief; I am not sure that the dogs had the monopoly of depression.

* * * * *

"In my hell it would always blow a gale," said the late R. L. Stevenson, who had the sensitive person's horror of a high wind.

But yesterday I sat at my window and revelled in Japanese effects. To-day a scorching wind is raging over the garden, withering the pansies and sweet peas, and making the casuarinas sigh sadly. The air is thick and yellow with dust, and I have taken refuge in the closed drawing-room, where the gloom is lightened by big bowls of poppies and wheat picked before the wind began its ravages. It was a cool wind this morning, and it will be cool, almost cold, again this evening, when the sun has gone down looking like a brass tray on the dusty horizon.

But the first day of Westwind leaves its grievous mark on the garden. The sweet peas go first; they cannot stand up against adversity, but with care the pansies will last awhile. Snapdragons and hollyhocks are brave, defiant things, and petunias, too, stay with us long, often until the rains break. Freshly watered they are odorous in the dusk, when we walk in the garden in the cool. Carnations, of course, flower their best in March, but if sheltered will go on flowering until May. There is a bunch to-day in the racing cup that is the pride of my silver table. Of course, what we grow here are not really carnations, only sweet-scented pinks, but they have the clove scent and a beautiful range of colours, and it is easy to make belief.

* * * * *

And so Spring is over in my garden. "Alas, that spring should vanish with the rose!" as Omar has said. I believe a time will come when not to be able to quote the bibulous Persian will rank as a rare and attractive accomplishment. But he was at least an ardent rose lover!

A LITTLE SQUARE BUNGALOW . . .
BESIDE A PEEPUL TREE.

CHAPTER V

My garden is mine no longer; in a new wilderness I am in process of planning a new garden. There is something sad in the mere words "A lost garden." I suppose the mind unconsciously reverts to the tragedy of our first parents; and then, too, a vast quantity of sentimental poetry has been written about deserted gardens, so that I could bring a whole host of pretty quotations to illustrate my position. But I am far too busy to do so. In the early years of my life I led, with my parents, a migratory existence. Like Jo, in 'Bleak House,' we were for ever "moving on." In contrast to those early years I have lived long and tranquilly in my wilderness; but, having been uprooted from it, I believe the roving blood of my ancestors speaks in me; the passion of the pioneer, stifling regrets, calls me to the instant making of a new garden, the formation of a new environment.

* * * * *

A strip of parched land beside a river bank, and a little square bungalow dumped down beside a big peepul tree, this is my new home. It makes me think of "Mariana in the South,"

> "A faint blue ridge upon the right,
> An empty river-bed before."

And there is the house also—

> "Close latticed to the brooding heat
> And silent in its dusty vines."

But I do not feel in the least, like either of the "Marianas" about whom Tennyson wrote—tiresome lachrymose young women, who would have worn away the patience of any man. It was an especial pity that the lady in the "moated grange" had no taste for gardening. She would have been so busy having the hedges clipped and the "blackest moss" weeded away from the flower plots that she would never have thought of saying "I would that I were dead," or have concerned herself very greatly over the fact that "He cometh not."

* * * * *

Someone has begun to make a garden here,

A WAVING SEA OF GOLD.

planting "duranta" hedges, and cutting square flower beds in the compound, which is iron hard at present. There are some stunted shrubberies of oleander and trumpet hibiscus, and some beds of canna. It is all orderly, and arid, and common-place, and my only present joys are a row of scarlet amaryllis in front of the house, and the lovely tangle of vividly green young sissoo trees that borders the compound on the river side. In the rains I shall send back to my old garden for plants and shrubs, and to France for roses, and shall be able to begin from the very beginning, planting shrubs in the order in which I have dreamt of them, massing colour and species. But alas! the rains are afar off as yet, and all I can do now is to draw plans that are invariably torn up on reflection. Talk of castles in the air! Gardens in the air are apt to be far finer, far more elusive, and more productive of despair and happiness.

* * * * *

At the bottom of the vegetable garden there is a group of mango and lichi trees, and amongst them, evidently planted by mistake, a young amaltas or Indian laburnum. There is something pathetic in the mistake that gave this beautiful, brilliant thing her wrong environment. She is

AMALTAS,
THE INDIAN LABURNUM.

like one of those non-workers one finds in the busiest colonies of this work-a-day world, and, like them, is probably treated with uncomprehending intolerance by those about her. To these useful, fruitful trees she must appear literally as a mere cumberer of the ground. To me she is a joy, as promising flowers for my house, of which I am sorely in need.

*   *   *   *   *

There are so many quick-growing and easily cultivated flowering shrubs in India that I never can understand why every garden is not full of them. Quisqualis is invaluable, and all the various kinds of gardenia flower in the driest and hottest weather. So, also, does poinciana, both the trees (gold mohur) and the shrub, while the lovely lagerstrœmia—the crape flower—is at its best in May and June. None of these are in the least rare or exotic, and yet there is not one of them in this garden, where English people have lived for years. There are so many ugly corners here that I shall require all the quickly growing things I can get. One of the most invaluable of shrubs is called *Belatee Chakoor* by the natives, and I do not know any other name for it. No one about there seems to know anything about it, but I shall get plants from my old garden. It has bushy dark

LAGERSTRŒMIA, THE CRAPE FLOWER.

green foliage, and flowers twice a year in big loose-petalled bunches of pastel yellow flowers. It is very handsome, and required no care at all but pruning, and I mean to fill many awkward corners with it. Then I shall have great clumps of *Tecoma stans*, also yellow; and, of course, acalypha and the whole hibiscus family—a large one which includes the beautiful shrub I called Persian rose for years, without in the least knowing why. In reality it is *Hibiscus mutabilis*, a coquettish plant whose blossoms are pure white in the morning, and pink at night.

Drawing-rooms are a good deal in my mind at present. Somehow, furniture that suited one room seems ungracious when required to adapt itself to the angles of another. And then, a move brings to light such a hideous accumulation of rubbish; such unexpected shabbinesses. I am seized with a desire to begin all over again, in my room as in my garden. Only so is it possible to have a colour scheme. My ideal room for India is green, with a good deal of white, and some discreet touches of mauve. Yet, as I write I remember that the pleasantest room I was ever in was yellow—so far as it could be said to have a colour scheme at all. The walls were a pale apricot tint, and there were yellow lamp shades and cushions and some big yellow flower-bowls. There was no silver table, that joy of the suburbs, and but few photographs. There were many books, flowers and plants in abundance, a big couch, and a cabinet with some old Worcester and Dresden china. The writing table held a sepia print of Rossetti's beautiful "Beata Beatrix," and on the walls there were some prints of good pictures, in broad plain wooden frames. Greuze's "Girl with an apple" was one, and Romney's portrait of Lady Hamilton as a Sybil was another. When I sat in that room I thought of the holocaust that my hostess must have had at some time of her life; for, of course, the first step towards having an artistic room is to have courage in *throwing away*. Stevenson said that a man who knew how to omit could make an Iliad of a daily paper, and there is no doubt that omission is a dominant note in Art. If you must keep old photographs they should be kept in decent obscurity. Break up meaningless ornaments, burn draggled hangings and shabby cushions. If a chair offend thee cast it out: only so will you have a room in which you can live and rest. If you cannot afford to replace these things go without; the average drawing-room would be vastly improved by a gap or two. Naked walls are not indecent, although some people seem to imagine they are, and if pictures are not beautiful and interesting in themselves no purpose is served by hanging them.

* * * * *

It is our dreary custom to make our writing tables into a family portrait gallery: to have presentments of Eric and Cyril and Marjorie in all stages of infancy; of papa and mamma foolishly studying a book of views; of aunt Edith and uncle Bob, and the baby. Could anything be more wearisome? I shrewdly suspect that the owner of the table is as bored by it as anyone else is; but all the women of her acquaintance have similarly adorned tables, and as proof of her affectionate domesticity, she feels it is bound to remain. And yet our domesticity can usually be taken for granted, while the question of taste needs emphasis. An artistically posed photograph is a delight when suitably framed and placed in comparative isolation. But the truth of the matter is that the average photo frame is a dreadful thing. We have got away from plush abominations, but we still have imitation leather, stamped linen, painted and embroidered linen, and thin embossed silver. I don't deny the fascination of the gleam of a silver frame here and there; but it must gleam, and, to put it crudely, I have seen too many frames that know no chamois duster.

* * * * *

It is the same with silver tables. If you have a few pieces of antique silver, they look well grouped on a dark wooden table. But scraps of cheap Indian ware, a couple of doubtful spoons, and a few modern frames, have a way of looking foolish when they are set out to attract and charm the eye.

* * * * *

If I was on the verge of becoming dictatorial on the subject of drawing-rooms, I ask forgiveness.

After all, it is only the small souls who make a fetish of their immediate surroundings, and the possession of a beautiful room is open to any vulgarian who can pay for the good taste of others. Personally, I have been absolutely happy in a room where plush monkeys scaled the art muslin draperies, and the chairs were enamelled blue and had ribbon bows on their legs. My hostess was sympathetic and gracious, and "'Tis the fine souls who serve us," even though they march in the ranks of the Philistines.

FIRST YEAR IN THE NEW GARDEN

E

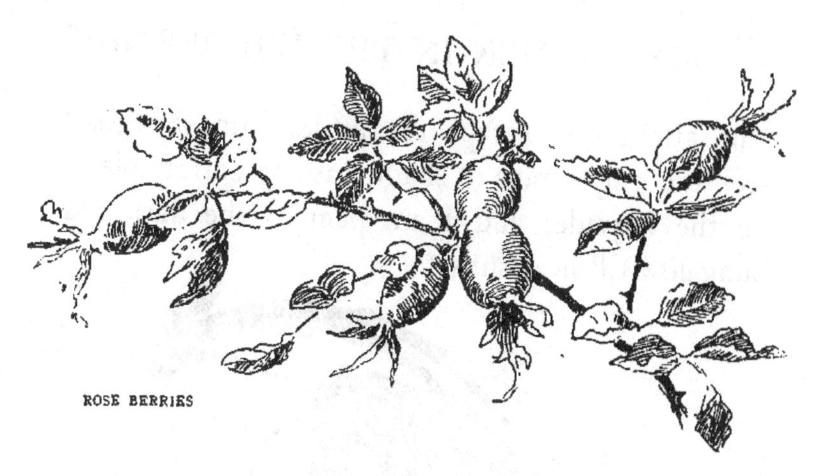

ROSE BERRIES

CHAPTER VI

MAY.

And, flaming downward over all,
From heat to heat the day decreased,
And slowly rounded to the east
The one black shadow from the wall.

TENNYSON.

MY chrysanthemum and carnation plants of last year, my palms and seedling shrubs have been arriving in cartloads, so the business of my new garden has begun in earnest, in spite of the heat. It is not auspicious weather in which to put them down, but I think the chrysanthemums will be comfortable enough under the group of lichi trees in the vegetable garden, and the carnations under the shelter of the eaves. Everything living seems to clamour for shelter just now: the village bullocks press hard against the mud walls of the huts, in vain attempt to escape the pitiless sun, the goats seek the tiny scraps of shade cast by cotton plants on the wayside, and I am pent in the darkened bungalow till near sunset.

THE LICHI.

I have been planting creepers against the house, quisqualis, and the brilliant *Bignonia venusta*, hoping that they will grow quickly and cover an ugly expanse of wall to the west. But, so far, they droop in melancholy fashion. The soil here is so very poor, and the big trees near the house sap so much of the good there is in it, that it is necessary to bring cartloads of richer soil from all parts of the estate, and I am growing more learned in the matter of soil than I ever thought it possible I could become.

Unfortunately, whenever a rose bed or shrubbery has been enriched with manure or leaf mould, the

THE TREES OF THE DRIVE.

little tendril-like roots of the trees, which are feeling all round, underneath the garden, say to themselves, "Hullo, here's luck!" and make straight for the patches of good soil, twining themselves about the roots of the plants, and nearly strangling them. The worst of it is that I shall never have courage to have any of the trees cut down; for it requires as much courage to cause a tree to be cut down as it does to perform any other irrevocable act.

* * * * *

I saw such a pretty garden not long ago, a little tiny scrap of a garden, with a piece of green velvet for a lawn and a shapely Indian laburnum in full bloom guarding the whole. It was all so neat and Dutch, with its closely clipped shrubs, and its rows of pot plants bordering the paths and lawn, and a prim garden seat beside a sundial, that I almost expected to see tulips in the borders, and a windmill somewhere on the landscape. It is the type of garden I have admired ever since I was a child and, in walks with my governess, used to pass and peep into a fascinating miniature demesne, where the tiny stream that fed a baby lake was spanned by a rustic bridge, and the summer-house and the lawn, and the flagstaff on it, and even the flag that flew from that were all on a scale suitable to the garden of a very large doll's house. But, admire it as I may, it is the kind of garden to which I never shall attain. I cannot drill my flowers.

With what is, I believe, considered a vulgar taste for ribbon borders I have never been able to indulge that taste, because my plants are not

ONE, WITH ORDERLY PARTERRES.

obedient enough to flower at the right time, nor in the right place; when my candytuft is coming out the eschscholtzia that I relied on for the other half of my yellow and white border has dropped its petals.

I have never gardened in England, but I have a fancy that the snug lobelias and calceolarias and geraniums are more manageable than any plants we grow in the Indian plains; and if that is so the people who tend window-boxes and public parks are not so unsurpassed in genius as I have hitherto imagined them to be.

* * * * *

I have an extremely inartistic love of tidiness in a garden. What I really would like is a little "boxy" garden, where the paths should be swept, and the leaves dusted every morning, and where no single petal of the blue ribbon border should encroach amongst the flowers of the yellow ribbon border. To check these tendencies, Fate, with its usual irony, has placed me in India, where gardens are by nature rebellious, and have to be appreciated in their own wild way or left alone altogether.

And so I appreciate—I appreciate very much indeed; but some day, when I go back to where there are

> "'Ouses both sides of the street
> And 'edges both sides of the road,"

we shall see what we shall see. Yet sometimes I fear that I, too, may say, with Kipling's trooper:

> "'Ow can I ever take on
> With awful old England again?"

and that the ribbon borders and the neat paths will seem trivial indeed after these "great spaces washed with sun." I wonder!

* * * * *

I have been looking through rose catalogues and compiling a list of those plants I want—or, to be exact, of those I mean to order, because I want them all. It is unnecessary that I should write my lists yet, for the plants should not arrive and be put down until the middle of November; but the compiling of the list seems in some indefinite way to bring the cold weather nearer, hastening the welcome day when we shall take down punkahs and put up mosquito nets, and lead for a season the normal British existence that is denied to us in the hot months. Then, too, the reading of the catalogues is a delightful occupation, abounding as they do in fascinating descriptions, such as: "Madame Philippe Rivoire: very large, globular; colour: apricot yellow; centre: nankin yellow; back of petals: carmine. Splendid." Or this: "Mrs. Harvey Thomas: elongated buds, oval shaped, coppery carmine; large flowers erect on long stalks; beautiful; very fresh and pure colour; sweet smelling."

My imagination is stirred by these gems of description, until I can almost see the fragrant things growing in my garden borders. The hybrid teas and hybrid perpetuals that I get from France have the merit of blossoming the first year they are put down; although, of course, their flowers are finer the second year. The teas are the most satisfactory, I find, for the perpetuals belie their name, and are not so free flowering as I expected them to be. Another thing that disappoints me a little is that there are not very many deep red roses amongst the new varieties. I love red roses with velvety black shadows in their petals, such as the old Black Prince and Monte Cristo. So far Etoile de France is the darkest rose I have had from France, and it is a lovely crimson hybrid tea, very free flowering; but I would like more varieties to mix with the yellow and pinks of the other roses.

> "Red rose alone is royal in her giving . . .
> And is no niggard, all her gold being spent;
> She gives her colour and her fragrance living,
> And being dead and dust, she gives her scent."

I suppose the Irish poetess who wrote those lines meant roses all, but to me the red roses always seem sweeter, richer and more generous, and I like to apply the words to them.

* * * * *

The other day I was asked for advice as to the

making of a garden, by some one who has no knowledge of gardening, and yet, living also in a wilderness, finds it imperative to cultivate one for the daily needs of fruit and flowers. To give advice is always so much easier than to refrain from giving advice, but it is with gardening as with life. Only our own mistakes can teach us,

THE GARDEN OF MY MEMORIES.

our own failures point the road to success. I am not one of nature's gardeners, only a flower-lover with ambitions, but from my years in a garden I have drawn a certain store of experience, which I am glad to proffer for what it may be worth.

One of my precepts is that shrubs and plants should always be massed; on the same principle that a few pieces of good china grouped give a far better effect in a room than the same pieces scattered. Plant in close lines or group your hibiscus, oleander, or moonflower, and you will have, at the right season, a splash of colour that will redeem your whole garden from the commonplace. Have clear spaces; don't mutilate your lawn with star or half-moon shaped beds which, unless you are an expert gardener, stand a chance of remaining empty half the year.

Except for leisured folk, annuals are best left practically alone at first. A mass of zinnias in the rains, a hedge of sweet peas in the cold weather, and some hollyhocks and petunias bordering the shrubberies; these are all I should attempt in the first year. For with a lawn and a shrubbery you have a garden that is tidy and manageable at all seasons of the year. I would have that kind of garden myself if I were not so weak, and did not yearly yield to the temptation of sowing pansies, carnations, and asters for the sheer delight of going out into the dewy cold weather mornings to see how the seeds are coming up.

If you contemplate making a garden you cannot learn too much about soils. *Malis* will stick a plant into any kind of soil and consider it an error of judgment on the part of the plant if it does not grow. It is quite useless to start any kind of pot culture until you have a large reserve of good leaf mould, and you must sift and mix your soils "with brains."

I believe I know why this business of garden-making attracts so many to it. It satisfies the creative faculty, that fever to produce, that delight in the work of our own hands, of our own brains, to which the ease of many lives, and sometimes life itself, has been sacrificed. Some women find this happiness in sewing, in seeing silken flowers and leaves take form beneath their fingers, the artist finds it at his easel, the writer at his desk. It takes many and diverse forms, but gardening is a joy that is open to all; for though we cannot all make pictures or books, the least gifted of us can make a garden. The growing of flowers calls for no especial genius or talent; unless genius be, indeed, a capacity for taking pains. Even the love of flowers, inborn in most people, can be cultivated, and it is barren soil indeed that with care and patience will not yield some good thing.

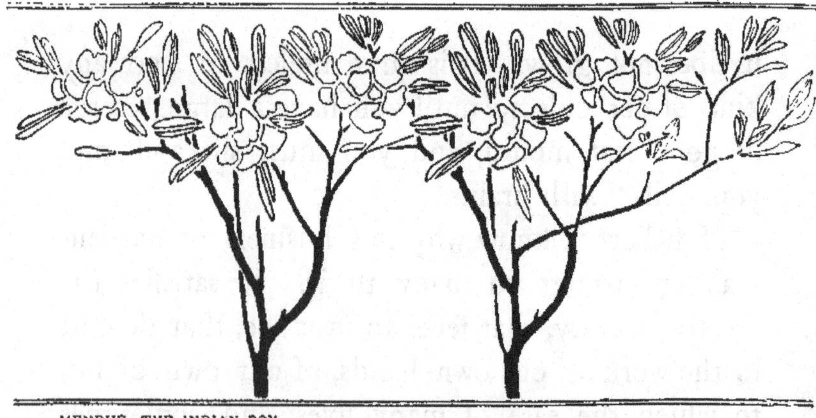

MENDHE, THE INDIAN BOX.

CHAPTER VII

JULY.

Give us courage, and gaiety, and the quiet mind.

R. L. STEVENSON.

Now that the rains have come, this garden takes on a certain beauty. Grass has sprouted in the bare patches on the lawn, the shrubs are branching, and the river has risen, so that we catch gleams of it between the sissoo trees that border the lower part of the compound. The ground drops sharply to the east of the bungalow, and on the cool breezy mornings one gets sometimes after rain, I like to sit there, watching the trees sway against the Sèvres blue sky, and the flight of the mango bird that haunts the little grove. Beyond the fence that encloses the trees are some sandy hillocks and a tangle of tamarisk bushes, while on the other side of the river there is a thick jungle of sugar cane, from the close cover of which jackals call out what the dogs take to be insulting remarks in the night. Sometimes during the pouring days we have had lately, they called in the daytime also, and then their cries sounded far more weird and unearthy than they do in the night, and were more deeply resented by the dogs.

BEYOND THE FENCE.

* * * * *

The horizon has been washed clean of dust now, and, in the evening, we can see long distances across the flat, moist country that is picturesque in its own unassuming fashion, with little groups of mud huts and some palm trees standing out against the clear sky. There have been some impressive sunsets, too, but I do not care much for these flaming skies; their gorgeous purples and golds oppress me a little. I like best the soft rose and grey and opal tints that follow rain in the cold weather.

We had an almost terrifying sunset last evening. We were driving at dusk along a high road that led westward through low rice lands. The sky before us was all soft gold and rose at first; presently the gold turned to flame, and gradually to scarlet. I never saw just that shade in the sky before—pure scarlet, with a heap of angry purple clouds piled on the horizon below it. The pools in the rice fields looked like pools of blood when the red light spreading over the dome of the sky, caught them; and as the darkness grew and the line of the road before us became indistinct, the flat land, and the red pools and the palm trees standing out against the fierce sky, gave the impression of some ugly dream. There came back to me something of what I used to feel when I was a small child looking at the Doré engravings in an edition of 'Paradise Lost' we had at home. The thunderous skies, the coiling reptiles, the angels and the devils, used to mix themselves into my dreams at night, but I was irresistibly drawn back to their contemplation in the daytime. Although it is years since I opened the book, I still can see the casting forth of the evil angels, as pictured in two awe-inspiring plates.

> "Nine days they fell; . . .
> Hell, at last, yawning, received them whole."

PETRÆA STAPELIÆ.

In the first there was the pell-mell of wicked souls, hurled forth from the "crystal wall of heaven," and falling through space in a hideous hurly of wings, limbs and armaments. And then there was pictured the open hell-mouth and the crushed and broken angels, pursued by lightning to the very floor of hell. No wonder my childish dreams were disturbed and a tangible hell became a more distinct possibility than it should be to the eight-year-old intelligence.

* * * * *

It is not usual to collect books in India, because most people are subject to moves, and the climate is supposed to be unfavourable to the preservation of anything more stimulating than a few colonial novels, and some magazines. But in my quiet years in the wilderness there gathered about me, almost imperceptibly, a small library that occasioned a certain amount of dismay when the question of packing arose. Then I gave away all I felt able to part with, but amongst those I kept was a box of bound magazines, 'Temple Bars,' 'Blackwoods,' 'Cornhills,' dating back to the sixties. On these I have been browsing of late days, while the rain has fallen steadily and gardening has been impossible, and from them I have drawn a greater store of interest than one usually gets from a magazine.

What pleased me most were 'Letters from India,'* written by the Hon. Emily Eden, a Governor's sister in 1835, a period of Indian history that does not usually arouse much interest now, coming as it does between two stirring epochs. Judged by her letters the Hon. Emily Eden was a simple and sprightly lady, with a pretty wit and a remarkably direct mode of expression. Letters *were* letters in those days, when two months was considered an abnormally short time for a mail to have taken between England and Calcutta, and the homeward letters were posted just when a ship was ready to convey them. Miss Eden was homesick, and when she got "ten fat letters" by one ship describes herself as "flumping" on her bed, and "wallowing" in her correspondence.

Apparently the Governor and his staff did not go to Simla in those days, but weathered it out in Calcutta, with occasional trips to Barrackpore, where Miss Eden speaks of riding on elephants in the park, and deplores the shabbiness and lack of comfort in the house, which, she says, is worse furnished than a London hotel!

A number of the Calcutta Society folk of that day objected to coming to Government House "At

* 'Up the Country.' Letters and Extracts from Hon. Miss Eden's Journal, describing the everyday life of Lord Auckland's Court.

Homes" at which there was dancing, and barred theatricals altogether. "On Thursday," writes Miss Eden, "we had a large levée; of those who did not come to the play, to show that they still visited us although we are so wicked; and of those who did come, to say they were much amused, and would still visit us, because we are so pleasant."

She asks her friends for books "to feed our poor yellow Indian minds," and she adds, "The more trash the better; trash is essential for India." She gardened and loved the fragrant Indian shrubs as I do, and built a little altar in her garden, which, not unfortunately, I should say, tumbled down in the rains. Of a lady of Garden Reach, who had managed to raise violets and sweetbriar, she infers that she must possess many other good qualities.

I like her references to her dog, Chance, "that little black angel," whom her attendants called "his little Excellency," and who curled himself into the wash-basin for security on a stormy day of their passage out. On hot nights Miss Eden took Chance under her mosquito house, to give him the benefit of the *punkah*.* I wonder much as to the mysterious person designated in the letters by a

* Fan.

dash, not even prefaced with an initial, which the others, at the least, get. I know that he would have preferred to be second potboy at the Pig and Whistle or sweeping a London crossing rather than accompanying the Governor's party to India, but I can make no guess at his identity.

Although she threatens often that she will cease to write, for she grows dull and a bore, I think her friends were to be envied in receiving these charming whimsical epistles with their glimpses of an engaging and sympathetic personality.

* * * *

I seem to have got far from the subject of my garden, possibly because the rain is too heavy to allow me to go out. It is coming down just now "like crystal rods stretching from heaven to earth," as Stevenson says in one of his letters from Samoa. It rained all night, but this morning it grew lighter for a while, and I spent a delightful hour pottering amongst my plants, with the soft, warm rain blowing into my face under an umbrella. I have had my potted chrysanthemums and coleus brought under shelter, and the verandahs look quite gay, for the coleus, from being in shady out-door places, is beginning to tinge with colour; I have pots and pots of it, and am propagating more to mix with my chrysanthemums later on.

I only hope all the squirrels and their families are snug in this downpour; I love the soft, dainty, graceful creatures. I have had pet squirrels, but the feelings of the dogs were so outraged by what they considered my unnatural and unsportsmanlike tastes, that I was compelled to give them up. Of course, no dog would have been sufficiently abandoned to molest my pets, but they signified their disapproval by dropped heads and averted gaze whenever the squirrels were obtruded on their notice, and my heart was touched by this obvious desire to avoid temptation. So when my squirrels were grown up I gave them their liberty, and they took up residence in a deserted nest and kept house there, and used to come down to eat the meals I took them, keeping alert eyes on my attendant dogs, and occasionally addressing abusive remarks to them; for squirrels are terrible scolds. I have seen them, at safe distances, shaking with the violence of the abuse they were heaping on infuriated dogs at the foot of the tree.

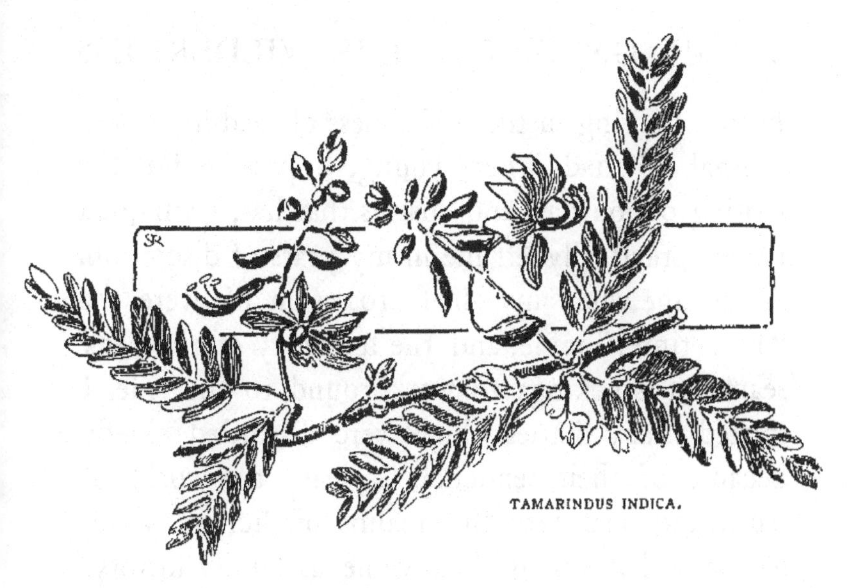

TAMARINDUS INDICA.

CHAPTER VIII

AUGUST.

In deep wet ways, by grey old gardens
Fed with sharp spring the sweet fruit hardens.

SWINBURNE.

It is raining steadily; the first heavy rain we have had for weeks, although the skies have been dull and heavy, and the sun has been badly missed. These grey, close days when it does not rain have a singularly depressing effect. We cannot do without the sun in India. Greyness is no part of the scheme, and the colourless days that we accept as part of the atmosphere of our own land are sad and dreary out here.

As gardening is out of the question, I have

been indulging in the foolishness of reading an old journal. I had, when young, a passion for the writing of journals and diaries that has, I am glad to say, practically left me in my years of discretion. At an age when my chief crosses in life were the "nine times" table, and the tendency of the back seams of stockings to work round to the side, I started many diaries which were discarded chiefly because of their tendency to leave the world of truth and soar into the realms of fiction, which naturally discounted their value as family history. At that tender age I could not, Macaulay-wise, rely on the vigour and purity of my style to excuse my inaccuracies.

If journals must be kept (and I suppose many of us go through a phase of this form of egotism) they should never be read—certainly not in this generation. Their self-consciousness and their crude attempts at self-analysis render them useless as human documents, and the only really interesting thing about them is their omissions, which occasionally are pathetic. Letters with their unconscious self-revelations are the only real human documents.

My discarded journal, from between whose pages slip letters, faded photos, newspaper cuttings, receipts, poems and sketches—all matters of much moment to the young—has yielded me one item of interest; a review of some letters of R. L. Stevenson's, five in number, written to a rising young artist; they are included in no collection of his letters that I know of. A limited edition was printed in New York, of which but two copies came to England, and from one of these the extracts given were taken. It is hard to think that not all that "R. L. S." wrote is for the public now. He was so full of strength and truth in all he wrote and did that we want all we can get of him. "We are not meant to be good in this world," he says in one letter, "but to try to be, and fail, and keep on trying, and when we get a cake to say 'Thank God,' and when we get a buffet to say, 'Just so: well hit.'"

"See the good in other people's work," he says elsewhere, "it will never be yours. See the bad in your own and don't cry about it; it will be there always. Try to use your faults; at any rate, use your knowledge of them."

"Pity sick children, and the individual poor man, not the mass . . . . and never pity fools!"

I do not know that we can, most of us, agree not to pity fools; it is such a pitiable thing to have ears that hear not, and eyes that see not, to be dulled to much that is beautiful or great in the world, to hear no music and think no thought—for so it is with fools; and—we have Carlyle's savage word for it that we are "mostly fools!"

The dogs are resigned to the weather, but they are not happy. They are crouching in an absurd fashion on the verandah, and looking out with envious eyes at the ducks, who appear to be finding a considerable amount of nourishment amongst the long grass, in the intervals of quacking duets with the "dhaboose" frogs in the pond on the other side of the hedge.

I am afraid that the dachshund's accomplishments of "begging," "trusting" and the like are having a bad effect upon her figure, which is deteriorating and taking on a distinctly "middle-aged" appearance, distressing in one so young. Yet it is not entirely greed that induces her to beg; she has formed a habit of employing this fascinating device for the furtherance of all her desires, and it becomes in turn a request to be taken into the dog-cart, to be played with, to be put to bed, to have doors opened or shut, and I have even known her to sit

THE FASCINATING DEVICE.

under a tree and beg for a squirrel. Her manners are indeed her strong point. Only yesterday I came upon her in the act of burying a bun, for which she had no appetite, but which she was too polite to refuse. When detected she looked apologetic and contritely endeavoured to eat the earthy fragment. On being reassured she retreated to a distant shrubbery and re-interred it. She wore a chastened air for the rest of the evening, and I have no doubt that the stodgy morsel weighed heavily on her, in more senses than one.

There are some dogs who have, like some humans, found the secret of perpetual youth. It is so with my terrier, who has lately been nicknamed Peter-Pan, because it is apparent that he will never grow old. In years he is old, so old that I am thankful for every year that passes and finds him still with me, but his heart is that of a puppy. Although his eyes are now dim and his little scarred face speaks plainly of many glorious battles, he yet can frivol on the lawn; can still attack a stalk of canna and lay it low; can rush it, and worry it, and tear it limb from limb, and hurl it high in the air, and roll on it, and leave it for dead and trot in with his head held high, to tell of his valiant deeds. He lives to himself, my little dog; he treats the other dogs with a certain consideration, but they mean nothing to him. There is, so far as he is concerned, only one person in his world, and when his little tired body rests at his mistress's feet, he tastes of such happiness as it is possible few humans know. For Maeterlinck, in his exquisite booklet, 'My Dog,' points out that a dog is, in a sense, more blessed than man, for his God is always with him—a tangible being, whose laws and desires are easy to understand, and whose anger or approval are swift and certain.

* * * * *

The only other dog of this household is a large shambling person whom, for want of a better title, we designate a greyhound. There is a fiction extant that she is not a house-dog; which, in reality, means that she has no especial chair of her own, and that at night she is chained up, while the other dogs roam at will. She is a gentle, melancholy creature, and her demeanour being that of a respectable charwoman she has been nicknamed the "Shabby Widow." As though conscious of her humble origin she has cultivated a spurious air of refinement. She never snatches or gobbles as better bred dogs do, and she affects a habit of delicately crossing her front paws when in repose that is, to say the least of it, misleading.

* * * * *

At times I find it difficult to believe that this is the place to which I came at a time of drought, of scorching winds and whirling dust. It is a bower of greenery now that the teak trees are in blossom, and a giant tamarind on the other side of the hedge shuts out the sight of the ugly buildings that depressed me so much. Even my new shrubberies are looking more hopeful, and I am rapidly becoming unpopular by begging plants with which to fill them.

Now that I have managed to get into my

THE "SHABBY WIDOW."

garden most of the common shrubs—tecoma and quisqualis, and the varieties of hibiscus—I mean to strive after a few that are more exotic. I have just learnt the name of a lovely thing that I saw rioting on a gate post the other day—*Petræa stapeliæ*—and of which I certainly mean to procure a plant this rains "as ever is." It is of a heavenly mauve, its long drooping sprays suggestive of wistaria, and although it is an ephemeral creature, with blossoms that one day's west wind will wither to naught, it is worth cultivating for the sake of those few days in the year when it may flower. I think it would look rather grand on a species of arbour that I am having constructed to the north of the house, with the object of disguising in some wise a horrid slab of masonry called a *chabutra*,* on which we are supposed to have our chairs placed in the evening. I prefer to have them on the grass in front of the house, and when the rustic trellis over the *chabutra* is finished it should serve very well as a nursery for my verandah plants, and the few ferns and mosses I have managed to cultivate without a fern house. There will be creepers over the trellis when I have dug up the hard and objectless path that surrounds the masonry and have put in some leaf mould and

* A platform.

THAT HORRID SLAB OF MASONRY.

good earth. And there will be more matter accomplished in my garden, which progresses very slowly—or so it seems to me, who watch.

* * * * *

Few Indian gardens are more fascinating than those attached to some of the older Behar factories; great rambling places, shaded in with fruit-trees, and odorous with giant flowering shrubs, amidst the rank luxuriance of which a garden lover may wander a whole morning of delight and envy. Such shrubs as are to be found here! tangles of scarlet flowering pomegranate, hedges of hibiscus, salmon pink, scarlet and cream, lagerstrœmia in

THE BIG TREES NEAR THE HOUSE.

all shades, chumpa with its waxen blossoms, fallen thickly underfoot, and its branches arching overhead. Gnarled mango and lichi trees there are, besides quaint fruit that go to the making of unusual preserves—*Aroo Bokhara*, the Indian damson, *Amra*, a hard astringent fruit, and what the *mali* calls "sapootas"—an insipid species of medlar.

I brought rich treasure-trove from a garden of this kind that I visited lately; white and pink Sandwich Island creeper for the arbour to the north. The pink variety is growing and climbing

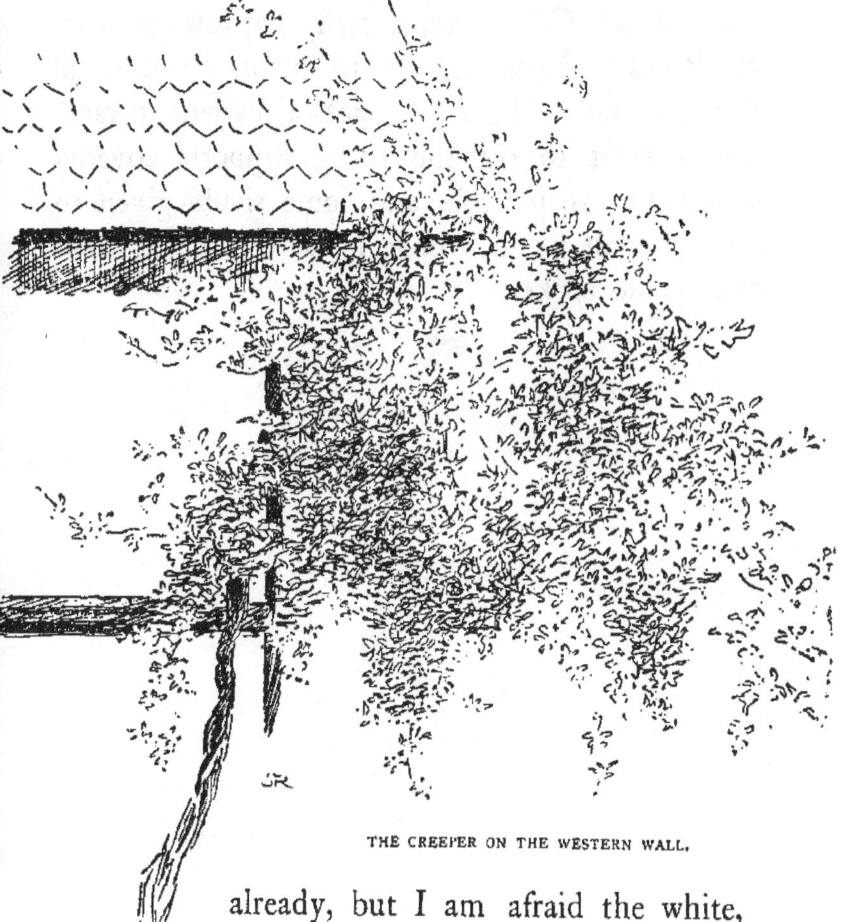

THE CREEPER ON THE WESTERN WALL.

already, but I am afraid the white, which I am told is a "difficult" plant, is not going to live. Thunbergia and *Tecoma grandiflora* I brought too, both heavy creepers, so that I'm afraid it will not be long before I shall have to thin out, and tear away many of the plants. Of course, a really capable gardener would have been content to plant a couple of creepers only, but I am not a patient person, and I want to see the rustic supports covered as quickly as possible and some shade given to my verandah plants, which I put outside as soon as the rains broke.

CARNATIONS.

CHAPTER IX

OCTOBER.

Scent of smoke in the evening,
Smell of rain in the night,
The hours, the days, the seasons,
Order our souls aright.

R. KIPLING.

HOT as the days still are, the knowledge of the coming autumn is inspiriting. When the smoke wreaths hang low over the fields in the evening then is the cold weather at hand; with the first chilly nights comes a sense of exhilaration such as belongs to the spring in England. Beautiful as is the English autumn it is sad, full of the scent of decay, and the threat of winter; but in India the season is a busy and a hopeful one. The rice crops are ripening, tobacco is planted out, mustard and linseed are springing in the lighter lands. With the sowing of the English flower seeds begins a new era, in which the discomforts of the past months are forgotten, and brilliant nights and cool sunlit days help to foster in us fresh courage with which to face a new year.

The evenings are drawing in; it is dark soon after six o'clock, and with so much to be done in the garden I grudge every moment that I am compelled to spend away from it. I cannot enjoy knocking tennis balls over a net while my mind is occupied with the seeds that await sowing and the seedlings that are ready to plant out; nor talk over prospective cold-weather gaieties with my thoughts running on unweeded paths and the proper clipping of hedges. I do not like this attitude of mind in myself, but it is there, and must be fought. For I always think it a pity when the care of a garden, or the love of books, or of home, or of any of the quiet, happy things of life, conduce—as they so frequently do—towards unsociability and the ignoring of social claims. There are really very few things for which it is worth while to renounce the society of our fellow-men. Every acquaintance is a potential friend, and I think we all recognise that, without friends, life would be a poor and empty business indeed. A friend of mine, a master at an English public school, once said to me that he considered no pleasure in life could compare with

THE COLD WEATHER IS OVER.

that of a talk with some person whose tastes—especially in literature—are akin to your own. Books are certainly wonderful promoters of intimacy, and can transform a casual acquaintance into a friend in the course of an hour. It is almost worth while going to parties that bore you on the chance of meeting some man or woman with whom you may discover mutual friends in that wonderful world that is peopled by the imagination of genius. Personally I would admit to my friendship anyone who reads Dickens with sympathy and understanding, and really appreciated dachshunds. For a kindred taste in dogs is another bond of friendship, and the dachshund—possibly because he offers such an irresistible challenge to the caricaturist—seems to appeal to the few.

It has always seemed to me rather pathetic that a dog so wise, so gentle, so reserved and yet so anxious to please, should have been given so grotesque an appearance; that his meek intelligent eyes and wistful expression should provide weapons to be used against him in the provocation of ribald mirth. Dachshunds take themselves seriously, but those who have kept them know their courage and affection, their unswerving obedience, and their faith.

* * * * *

There is, mercifully, no season in India when all the trees are bare; at present the cork tree is scenting the air with its drooping bunches of starry white flowers. But with the fall of the teak and sissoo leaves the homesick heart turns yearningly to the thought of the fall o' the year in other lands. Mine is stirred with the memory of russet tints in Perthshire woods, the scarlet of the wild cherry trees, the gold of bracken, and the fading purple of the heather on great Ben-e-Vrackie. Ah me! the Atholl country! the glory of early autumn in the Killiecrankie pass, the greeny-brown peat shadows in the eddying waters under the Cluny Bridge, the great stretch of moors where

> "Blows the wind to-day, and the sun
> And the rain are flying . . .
> My heart remembers how!"

There is no doubt that the Highlands cling; the Highlander never forgets his homeland, and I am bound to acknowledge that he, not infrequently, makes himself ridiculous in the eyes of the mere Sassenach by the somewhat aggressive nature of his enthusiasms. "Heather for Scots abroad" is taken quite seriously by a well-known Scottish weekly, although I fancy it will be some time before the *Daily Mail* takes up the question of providing Englishmen in exile with oak leaves, or acorns, or bunches of lavender from cottage gardens.

* * * * *

Being an extremely conservative person I am endeavouring to make my new garden look as

much like my old garden as it is possible. To this end, I am at present having rustic arches erected to span the path that leads from the house. I shall have to train nasturtiums and morning glory over the arches for this year, until the roses I have sent for are ready to climb. On either side of the path I have put a border of chrysanthemums, and on the south side of this again I shall cut a wide border for roses. The north side must be left alone, for it is too near the trees of the drive, and roses demand sun. There are only four rose bushes in this garden at present, and they are in so shady a spot that they will have to be transplanted directly the weather is cool enough. The better to recall my old home I have cut borders round the house under the eaves, sowing nasturtiums on the nearer side, to train along the edge of the verandah, and on strings to the roof. In these borders I have planted coleus which should colour vividly with the first chilly nights, and on the north and east sides I mean to plant violets in the middle or end of October. Unfortunately the house faces west, so that no violets can be grown there. The hanging baskets and rustic stands I have had made for my verandah are replicas of those I had in my old home; and, though I have not yet built a fern house, I certainly hope to see one in the shade of the big peepul tree before many months are gone.

* * * * *

A great deal of thought and effort and hard work has gone into my garden. Yet there are days when the contemplation of it brings nothing but depression, when I seem to have worked hard and achieved little; when the shrubberies look poor and ill-arranged, the grass full of weeds, the pot plants sickly, and the whole scheme of things a mistaken one. All gardeners know these moments; without them and the incentive they give to further effort and more determined progress, I fancy there would be but few beautiful gardens. For the true gardener, like the poet, is touched with divine discontent.

THE TWO-HANDLED CHOCOLATE CUP.

CHAPTER X

NOVEMBER.

Wherever men can live healthily, bright colour is given to them in sky, sea, flowers, and living creatures. RUSKIN.

SOME one has truly said that the chief pleasure of going away from home is in coming back again. I have been away for ten days, and my garden, in the delightful way Indian gardens have when released from their owner's too anxious care, has been preparing surprises for me. When I left, the chrysanthemums were budding in a sulky fashion, apparently determined not to hurry themselves about going further, the poinsettias were but just forming their pointed red tongues, and the English flower seeds were germinating in a distinctly unsatisfactory manner. I returned to find the garden in a riot of colour—a few cold nights have painted the tips of the acalypha shrubs a vivid crimson, the poinsettia is ablaze, and the yellow chakoor with which I filled every available corner is heavy with blossom. The misty blue of the morning glory that I have set on stands in the new shrubbery, and the heavy crimson of a line of cockscombs that border it, complete the scheme of colour—except for the chrysanthemums. They are there in all shades, and in every variety, thick lines of them on either side of the path that cuts my lawn, pots of them on the verandah steps, and sunk amidst the coleus that I planted under the eaves, and which the cold has also touched to scarlet. Friends were kind when I was making this new garden, and sent me many chrysanthemum plants, and, as I had not the least idea of the varieties, I planted them all haphazard, with a very fine effect, I think. Some, which I put on the north side of the house are not flowering, as they have not enough sun, but I shall profit by my mistake, next year.

I miss roses at this season; or shall when the chrysanthemums are gone, and there comes the dearth in Behar gardens that only roses can fill. For some reason of his own my old French rose-

grower has not yet sent me my rose plants; European growers are so fussy about getting frost to the roots of their roses that I fancy that has something to do with the delay, which is none the less vexatious. I can only be thankful that cosmos will provide me with flowers for the house until the

English flowers are abloom in February. I have sown it in every available corner this year, and feel that I never fully appreciated its worth until I saw it all a-riot in Darjeeling. Red-roofed Darjeeling was a dream of cosmos in every nook and cranny; knee-high, waist-high, neck-high it grew, a glory of mauves and white. Looking up the hill I saw cosmos awave against the blue; it hung against the indigo mists of the valleys, and flaunted itself against the dazzling background of the snows. I thought I had seen the best aspect of Darjeeling when I saw it aflush with rhododendrons in the spring, but, smothered in starry cosmos, with glorious relief of dahlias, I feel that I can hope to see few more beautiful places.

* * * * *

Darjeeling fired me with ambition to grow cactus dahlias, which are, I think, among the most beautiful flowers I have ever seen. I am assured by an authority that there is no reason why I should not grow them in the plains, and there certainly is no reason why I should not attempt to do so. I have just received some tubers, and am planting them in pots of light soil—leaf mould, with a sprinkling of sand and charcoal, and a little old sifted manure. I hope I am not making any kind of mistake, and that the end of next rains will

see my cactus dahlias in bloom. I shall be a very conceited person when they do flower.

* * * * *

Carnations are among other plants that I shall have to care for through the rains. The first lot of seeds sown this year failed, and though the second lot have germinated thickly I fear that they will scarcely blossom this season. I don't like the idea of going through a whole year without my favourite flower, but the change of abode or the heavy rains we have had killed all my last year's plants. I think the probable reason was that I was not well enough acquainted with my new garden to find the exact spot to suit them in damp weather. There is nothing reasonable about carnations or violets—given every condition under which gardening books tell us they should thrive, they frequently fail to do so, and then will flourish amazingly in a quite unlikely spot. In my old garden I had discovered the exact places that suited my carnation plants at all seasons, but I still have much to learn of this garden. I am inclined to fix three years as the period it takes to really know a garden, and an untried garden is like an untried friend—attractively full of possibilities.

I am so glad that I followed out my theory of massing my various shrubs instead of planting them in lines; at present one end of my shrubbery glows with poinsettia, and the other with bronze-leafed castor-oil plant, and chakoor, each splash of colour being a delight to the eye instead of a tantalisation, as it would be if the colour were dabbed in. I love colour; vivid tints under the sun, soft tints for the shadow; to choose colour, and to know how to manage it is a God-sent gift. What a different place the world would be if every woman were born with a sense of colour, and would consent to clothe herself in tints of the same colour instead of attempting incongruous mixtures at the dictates of fashion. Think of the effect of a room full of women each clothed in one tint, or in shades of the same colour; the amethyst woman, the woman in blue, in brown, in yellow, in grey, in black—no mottled effects; all pure colour. But of course, such a dream is too good to be true.

I live every moment of the day just now—a lamentably short day when the sun rises at seven and sets soon after five o'clock. There is not very much to be done in any garden at present; some English flower seedlings to be bedded out, and a lot of watering, which the coolies do mechanically. But there is a great deal to be enjoyed; the misty early mornings, drenched in silver dew; the brilliant golden noonday that can yet be ventured out into; the chilly opal-tinted evenings, and the long, quiet hours after the lamps come, while the firelight flickers over the old china on the cabinet and makes the dachshund's coat shine like satin as she lies blinking on the fender stool.

Like Charles Lamb, I have a partiality for old china, and, like him, I am not conscious of the

BLINKING ON THE FENDER STOOL.

time when china jars and saucers were introduced into my imagination. I cannot, however, pretend that I always appreciated them; in my rebellious youth the dusting of the precious pieces was a duty performed in sulky obedience, where now it is a labour of love. For no servant is permitted to touch the two-handled chocolate cup, made two hundred years ago, for the Emperor Augustus, the Sèvres casket of heaven's blue, the Worcester vases, or the pepper castors and salt-cellars of old Wedgwood—not the Wedgwood most of us know of classic figures on dark blue or green, but the old "fruit basket" pattern, with dappled-blue edge. China is not a usual possession in India, and mine, secluded in its corner, passes, for the most part, unnoticed. It is travelled china, having been over half the world even in my remembrance, and it seems to have found fitting harbour, at last, in this quiet home.

* * * * *

In these quiet evenings by the fireside I have awakened to a disappointment. Years ago I loved and possessed a certain book; no deep or abstruse work, just a novel, of which the characters were sympathetic to me, and the setting quaintly picturesque. I returned to its pages again and again, and always with enjoyment. Then I lent it. I have a habit of lending my books; a foolish and mistaken habit, and, ingrain, a selfish one; because one does not usually lend a book purely for the sake of giving pleasure to the borrower, but for the sake of probing a mind, and gaining an opinion. Of course the habit is set round with the disappointment it deserves. True book-lovers seldom borrow books, any more than they read the Library of Famous Literature or volumes of that ilk. They long to possess, and a book that they cannot return to is a mere aggravation of the intellect. And if you lend your books to any but the real book-lover he cannot understand why you are inclined to make a fuss over the loss of a paltry cloth-bound volume that he has left in a railway carriage. My book was lost four years ago, but it was only a few months ago that I sent to England for another copy. And now I am sadly realising that the old glamour is gone, and that my book friend charms me no longer in the old way. As literature I still enjoy the delicate style, but there is no longer magic in the pages, and I feel as though I had lost a human friend. I try to hope that this is merely a misunderstanding, and that, in some other mood, I shall find my friend again, but it is true that we bring only from a book what we take to it, and if I fail to find the old thrill and delight I fear it is because I am now cold and dull.

SECOND YEAR IN THE GARDEN

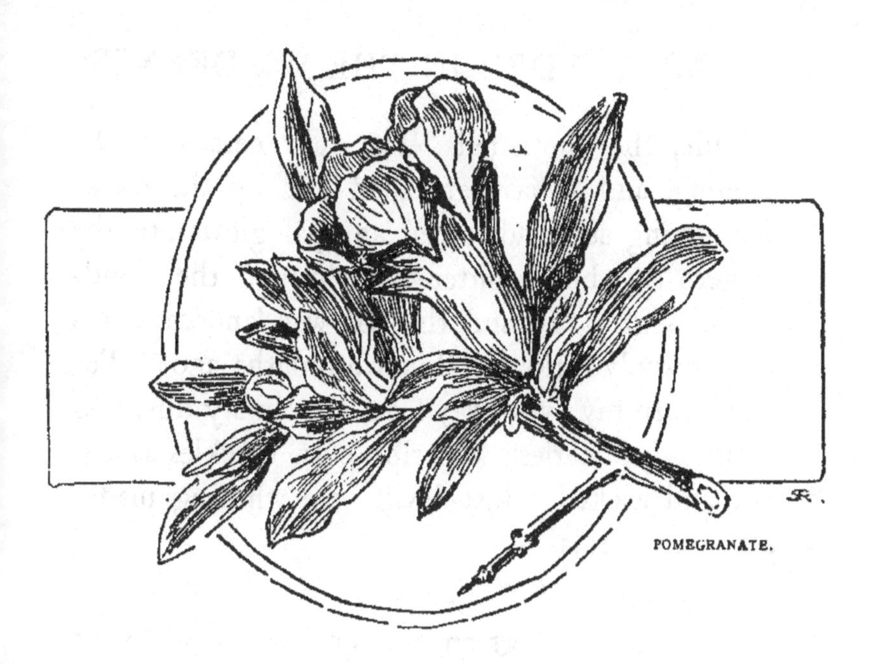
POMEGRANATE.

CHAPTER XI

APRIL.

When the last of the roses
  Despairingly closes,
In the lull that reposes
  E'er storm-winds wax fleet.
A. L. GORDON.

Now that the cold weather is surely over, and the English flowers are withering before the rushes of hot wind from the west, it is difficult to recall any other condition of climate. It seems that somewhere in the dark ages occurred the rainy January day on which I closed the doors against the wind, and sat by the fire pretending to myself that, outside, the snow fell. The rain ceased in the evening, and I was persuaded out by the dogs; the setting sun had forced an opal glow into the western sky, but a bitter wind hurried the clouds across a grey sky, and the flat wet landscape was inexpressibly forlorn. But two months ago! To-day I turn my eyes away from the thirsty dustiness of my shrubberies; the ripe corn crackles as it is cut, and wicked "dust-devils" are whirling madly across the fields.

* * * * *

The garden is green enough, for the watercart creaks about it all day long, prolonging the life of the phlox and coreopsis. The eschscholtzia has gone, and the grass is strewn with poppy petals, but the late sown carnations are coming into flower, and there are some pots of flaming portulaca on the steps. Because the house faces West, and portulaca must of course have sun, these are on the back verandah steps, but that does not concern me greatly. Although my guests do not see them I know that they are there, and it does not seem quite fair to flowers that they should be used only to create an impression on your acquaintances. The roses are at the back of the house, too, because a small piece of sward there seemed the only possible position for them. There are more than a hundred new plants, put down in December, which was, of course, far too late; but it does not seem to have harmed them at all, and they are looking very cosy and green, and are even blossoming—little delicate blooms, obligingly put forth to let us know their possibilities in the matter of colour. The man of the house considers the roses as his special province, and, as the dachshund does not believe it possible that he can exist for a single hour without her society, she also wanders amongst the tiny bushes with the sapient air that comes natural to her breed. She is a restful creature, now

BANYAN TREE.

that she has outgrown the ardours of puppyhood, and it is characteristic of her that she should have, voluntarily, attached herself to the one member of the household from whom she has received the least expression of affection. There appears to be a kind of silent communion between them, and an absolute content.

* * * * *

Although it is impossible not to regret the cold weather, yet the quiet days under the punkah bring their compensations. Life is a hurried business in the short crystal days of the Indian winter; the anxiety not to lose a moment of the brilliant weather grows positively feverish; the rush of pleasures; the gaieties, the coming and going of friends, the general human interest—this is absorbing. It is Life, but with no time for living. With the longer days comes in a different era; for me there are restful mornings in my shaded drawing room, time to think, to read, to sew, to write letters, or even to lie still on the big couch, scarcely thinking, and only vaguely appreciative of the glow of a vase of scarlet amaryllis against the faint green wall. Idling!—most sternly condemned in one's youth, grows to be almost a precious accomplishment in later years. When I have spent a futile morning in talking nonsense to the dogs, in reading passages from my favourite books, writing letters that never are to be sent, dipping into gardening books or playing patience, I read Stevenson's 'Apology for Idlers,' and feel happy and well content. Justification is a pleasant thing, and my youth was so strenuous a one that I cannot idle without feeling guilty, even before my servants!

* * * * *

I often think the present-day child has far too strenuous an up-bringing. The very word "strenuous" is over-worked: scarcely a conversation that does not include it. Parents are so anxious to impress on their offspring that they must never idle that by the time the childen are grown up they have lost the knack of doing so.

The public school boy has few "slacks," but then, of course, the theory is that the human boy is so constructed that unless he is overworked he will get into mischief, which is probably the case. The girl at home goes from music master to drawing lesson, from cookery class to "gym," from singing lesson to hockey; probably she has some deeper subjects on hand also. The luxury of an hour with an apple and a book in the long grass of the orchard, or the winter gloom of the garret is a surreptitious one, amounting almost to a crime. She has no time to read, other than the subjects that comprise her "work"; no time to sew—that would interfere with her studies—nor for the "long, long thoughts" of youth.

The result is that women, having forgotten how to idle, take themselves too seriously. It is almost tragic to think of a time when the merely sweet, merely frivolous woman will have ceased to exist, and only earnest minded Christian scientists and suffragettes will be left. After all, the woman who knows little more than how to do her hair has excellent chances against the most erudite of her sex. Women of this type have helped to make history.

It is refreshing to be, sometimes, in the society of quite foolish persons; people who have no earnest convictions, no opinions—best of all, no information; people who gossip, and laugh over novels, and make jokes, and ask riddles. Life is an anxious business for most of us, and the few who can preserve some lightness of heart are to be envied and cultivated. But little leisure was recognised in my own youth; yet it held some quiet, happy times, when I snatched at happiness and found it where I never have since failed to come near to it—in reading. I think, sometimes, of the busy hours in the dim dairy, so far from the house that I started at every sound or shadow, and breathed freely only when I could store away the butter and run to the house and cheerful company. Then the long seams to sew, the endless socks to darn on hot afternoons when "the blue fly buzzed in the pane," and outside the blue gum trees stood straight and tall against the pitiless sky, and the air was sickly with the scent of wattle-blossoms. Against these weary times are set memories of hours in the quiet room where books were kept, and where, if you sat very still on the floor by the bookcase when the door opened and your name was called, you might escape observation. Poetry I read in quantity. Longfellow, Byron, Moore and Hood, Shakespeare's tragedies and some of his comedies, Ouida and Dickens, many old volumes of 'Temple Bar' and 'Cornhill' and unlimited 'Family Herald.' No one forbade me anything; no one told me what to read; I think I must have suffered from slight mental indigestion, but there was no really unwholesome literature in the house, and I do not think I sustained any real hurt. Later, I, of course, "discovered" Browning, Rossetti, Tennyson, and Swinburne, and just the other day a friend sent me the early 'Romances' of William Morris, with which I am at present enchanted. To me they seem the poetic expression of Burne-Jones' pictures. Mediæval maidens, rose-wreathed and clad in scarlet or in russet, bid farewell to lover-knights in walled gardens of "that happy poplar land," where:

> "The hot sun bit the garden beds
> *When the Sword went out to sea;*
> Beneath an apple tree our heads
> Stretched out towards the sea;
> Grey gleam'd the thirsty castle leads
> *When the Sword went out to sea.*"

There you have a Burne-Jones canvas complete! The maidens were adorably named with such names as "Ellayne de Violet" and "Constance fille de fay"; but their knights were seldom faithful, it seems, and the ladies wept much and openly, and sang such sweet songs as these:

> "Gold wings from sea to sea
> Grey light from tree to tree,
> Gold hair beside my knee
> I pray thee come to me,
> Gold wings!"

Then they died of love, for hopeless love killed in those days, when passionate gold-haired women had naught to do save "broider shields and sigh for their lovers."

QUISQUALIS IS INVALUABLE.

CHAPTER XII

MAY.

And a bird to the right sang "Follow!"
And a bird to the left called "Here!"
And the arch of the woods was hollow,
And the meaning of May was clear.

SWINBURNE.

I AM envious and discontented because I have no gold mohur trees, *Poinciana regia*, in my garden. I planted two last rains, but I shall have to wait a year or two before they blossom, and in the meantime I am jealous of the specially lovely appearance of other people's trees this year. I have just been staying at a house that is set on the side of a lake and closed in by these trees. I slept in the verandah that overlooked the lake, and opened my eyes every morning to a gorgeous sight of flame-coloured blossoms closely massed on nearly leafless branches. Beneath these branches I could catch the silver ripple of the water, above which the plover flew and circled, asking their eternal question: "Did he do it?—did he—did he—did he do it?"

One night I staggered sleepily out and saw a comet, a golden streak on the dull blue sky, and tried to feel suitably impressed, and felt only sleepy, and rather cross, and chilled by the east wind. I have a horrid disposition to admire such things as appeal to me personally, and not the things I am told I ought to admire. I remember, even now, my boredom when, at seven years old, I was taken to a Handel recital and told I ought to enjoy and remember it. All I really remember is that the lady who sat in front of me had a white tacking thread left in her mantle, and that I ate chocolate creams on my way home. I must have been a horrid, stolid, mundane child.

* * * * *

Kipling tells us somewhere that

> "Scents are surer than sights or sounds
> To make your heart-strings creak,"

but I think that the sounds of birds are just as effective for the painting of vivid memory pictures. I shall never now hear the plover cry without seeing rippling water, and the branches of poinciana trees against the pale dawn sky; just as the harsh cry of the blue jay means sun-dried lawns, and gardenia flowers, and the chatter of green parrots brings back my old garden, and a mass of scarlet hibiscus blossoms scattered on the path. There is a bird,

ON THE SIDE OF A LAKE.

too,—I think it is the crow pheasant,—that says "honk, honk" continuously when rain is coming; the sound seems inseparable from soft, showery days in my old garden, hot moist days—"growing weather" as we used to say in childhood.

I pass over the horror that the cry of the "brain fever" bird brings to the newcomer to India; it was years before I could hear its cry with any complacency. Yet I had heard the laughing jackass in its native haunts, and neither that nor the screaming flights of sulphur-crested cockatoos were especially distasteful to me. Perhaps they fitted too well into their environment; the gaunt "wrung" trees in the paddock, the tangle of "ti trees" that grew so densely about the lagoon and yet gave no shade. What I think of oftenest in that scene are the bushes of double scarlet geranium that grew waist high against the house door. What would I not give for those bushes in my garden here, but in those days I thought of them as being almost too common to be beautiful.

* * * * *

The cooing of doves is associated with my earliest days in India; they built in the wide verandah of an old house that is now deserted, and

I know very well what Longfellow meant when he wrote of

> "The Sabbath sound, as of doves
> In quiet neighbourhoods."

They suggest quiet Sunday mornings in the country, the sound of church bells, and the peace of spring sunshine.

Blue jays and gardenias and the sickly scent of the sirus tree—I hate them all. Gardenias are dreadful things, suggestive of such uncomfortable functions as weddings and funerals; their scent is languorous and wicked, and their thick waxen petals are repellent to the senses. There is nothing real or flower-like about them; I infinitely prefer immortelles, which are, after all, but the mere travesty of a flower.

* * * * *

May in Behar is not the May of the poets; it is a long weary month, a month of dust-storms, and parching winds, and scorching sunshine. I have planted out the chrysanthemums in the shade and the carnations are under shelter, and now there is nothing to be done in my garden but to wait for the rains, and to keep life in the drooping shrubs. For flowers I have only the quisqualis that I have succeeded in training up the western wall of the house. My old garden had some shrubs that came bravely into flower just now—a kind of shrub, bauhinia was one, with a veined heart-shaped leaf, and a delicate azalea-like flower—and I think that being near a tank gave richness to the soil, for I do

THE SCORCHING SUNSHINE.

not remember that it ever looked such a wilderness as does this garden now.

* * * * *

Anxious as I am to acquire creepers for the rustic arbour to the north, I yet am afraid to introduce the ubiquitous passion flower, *Passiflora cærulea.* I remember too well a garden—the garden of the house of the doves—that became obsessed by passion flower. There was an archway leading to the vegetable garden, and in the beginning the creeper seemed a suitable covering for it. After a year the problem that presented itself to every member of the household was how to circumvent the aggressive creeper. The rooting out of the original plant helped the situation not at all. Rootlets sprang up everywhere; amongst the potatoes, in the shrubberies and drive: the paths crumbled with it, and there was even a legend that it was found sprouting from a damp wall. That garden is now no more, and possibly even the house is gone, but I have no doubt that the passion creeper continues its riotous career, and that the briar that surrounded the Sleeping Princess was not to be compared to it.

* * * * *

It is, perhaps, as well that there is little to be done in the garden just now, for the coolies appear to be suffering from a serious epidemic of marriage festivals. They are always going on leave, and coming back clad in yellow garments that suggest Chinese mourning. Pink evidently does not suggest itself as a festive colour for these occasions, for they practically never appear in that. These cloths are all dyed in the villages, with vegetable dyes, which are very clear and delicate, though not lasting. I wish that Indian marriages did not involve so much beating of tom-toms and blaring of conches. The drone of tom-toms in the thick, dust-laden air is one of the most depressing of sounds, and through the hot nights we have had lately they have beaten incessantly like the pulse of the night. If the coolies do not return to their normal state of mind by the time the rains break I shall be forced to do my own garden work, for I am forming huge plans for this year, and do not mean to be baffled. R. L. Stevenson, weeding sensitive plant on his Vailima estate, tells us that "Nothing is so interesting as weeding, clearing and path-making," and he adds that if he makes sixpence by plying his spade, his "idiot conscience" applauds him far more than if he sits in the house and makes twenty pounds by his pen. I love those letters written in the shadow of the Vaea Mountain by that great, simple-souled man who lived in every instant of his life, and yet told his friend that he was only happy once. Well, and to have been happy, and to have known it, is to have received a gift of the gods, for most of us recognise happiness only when it has departed.

THE UNDERLINGS IN MY GARDEN.

CHAPTER XIII

JULY.

Whenever the trees are crying aloud,
And ships are tossed at sea,
By, on the high-way, low and loud,
By, at the gallop goes he,
By, at the gallop he goes, and then
By he comes back at the gallop again.

R. L. STEVENSON.

A BLUSTERING wind has raged over my garden for two nights and two days, and I have been proportionately miserable. I can never say how much I hate a high wind. At night when it sweeps round the house the branches of the big peepul tree crack and groan, and the ceiling-cloths flap with the sound of minute guns; then even the dogs are restless and uneasy, and I lie sleepless, and at each fresh gust think of Stevenson's quaint fancy of the horseman coming and going in the night.

Now a fine drizzle has succeeded the wind, and I am out again, planning shrubberies on the graves of the removed teak trees, and trying to rectify as many of my last year's mistakes as is possible. Many of my shrubs were planted last year in unsuitable places, in shade where they should have had sunshine, or hidden by larger, quicker-growing shrubs. So I am getting them gradually into new places, and am filling up spaces with canna which likes new soil, and with the red-leafed variety of castor-oil plant, which seems to me invaluable in filling up a shrubbery.

My work lies much in the fern house at present, where I am putting out boxes and boxes of coleus cuttings, and watching the baby palms that are to be potted as soon as the rains come. I am absolutely convinced that if you are ambitious about verandah plants—as I am—you must renew them yearly; that is pot out, in the beginning of the rains, the cuttings that you struck last rains, and banish your last year's plants to a shady shrubbery. It seems such a very simple rule, but I see many untidy verandah plants and am so frequently asked what is wrong with them. Old age is their usual malady. Crotons may look well into their second year, with repotting, but dracænas and panax must be renewed, and my hanging baskets are also claiming my attention at present. The cyanopis, lycopodium and saxifrage with which they are filled are beginning to look droopy and weary, and I am going to replant them all. If I had been provident I would have had another set ready to take their places in the verandah, but I fear I am not a true enough gardener to look very far ahead.

I imagine that the rains, whether they come early or late, never fail to take gardeners by surprise. Of course there exist some prudent souls who planned and dug their new garden plots in May, and sowed their balsam seeds when the first cloud of the coming monsoon blew across the sky. But the average gardener is, by temperament, of the many who live in the moment, and it is the first really rainy day that serves to galvanise him into feverish activity. I have during the last week become extremely unpopular with the underlings in my garden, who have found themselves hustled and goaded and scolded into doing an appreciable amount of work. There is so much to be done that I feel as overwhelmed as the coolies do, with the exception that my dreams are haunted, as theirs certainly are not, with visions of all that still is left undone. There are beds to be weeded, hedges to be clipped, palms and ferns to be repotted, soil to be mixed for chrysanthemums, violets and carnations to be carefully guarded against damp, seeds to be sown, seedlings to be put out, shrubberies to be planned, shrubs to be planted—dear me, life in the wilderness is really a most strenuous affair! Gilbert Chesterton says somewhere that people really travel to get away from the intense excitement of life in their own backyards, and because staying at home is so very much more exhausting

than going abroad. The paradox is easy enough to grasp when your backyard happens to be an Indian garden, into which the exigencies of the

SOME ONE HAS BEGUN TO MAKE A GARDEN.

climate compel you to put a year's work in four fleeting months.

* * * * *

Besides balsams, amaranthus, and zinnias—a

curled and crested variety from which I hope great things—I have sown hollyhock and morning glory, and cosmos, for I am convinced I was too late in putting down these seeds last year. Early next month I mean to sow carnations and wall-flowers—the latter as an experiment, for I really have very little idea as to when they should go down, but when I sowed them in October they grew big and bushy but did not flower at all.

* * * * *

There is very little to be done for the roses just now, which is rather fortunate at so busy a time; just to keep them from the heaviest showers and from getting water-logged and to circumvent the white ants, if they show a disposition to attack. I gather a small bunch every day, and pay no attention to people who tell me the plants should not be allowed to flower out of season. The flowers have no decided form, of course, but their colours are promise of what I may expect in the cold weather, and have a delicacy I never seem to see in any roses but those I get from France.

I have spent some long, happy mornings under the trees in my potting corner; showery days when the grey clouds pile themselves on a dull blue sky, and the atmosphere is suggestive of a Turkish bath. I like the damp heat, so far as I can appreciate any heat at all; I breathe more freely now that the air is washed clear of dust and the evenings bring golden sunsets, in the stillness of which the trees in my garden take on the unreal appearance of trees seen in a stereoscope.

* * * * *

I am at present feeling far more hopeful about the gardens than I ever have felt since I came here. The rains in India work such magic; already the creepers have formed a delicate green lace-work up the sides of the arbour, and the western end of the house is almost hidden in *Bignonia venusta* and quisqualis. When, in November, the rose bushes have been shifted to the long border in front of the house, and the chrysanthemums and poinsettias are aflame, I believe I shall be almost satisfied—if gardeners ever are satisfied!

Now that the days have begun to shorten—such a very, very little, but still to shorten—there begins that pathetic "looking forward" that is the salient feature of our life in India. "When the cold weather comes," we say, and look at the calendar, and think how near October looks—now. And yet how they drag, those last months of the hot weather, and in September, how difficult to believe that cool October will soon be with us.

I AVOIDED THE VILLAGES.

CHAPTER XIV

SEPTEMBER.

I watch the green fields growing,
For reaping folk and sowing,
For harvest time and mowing,
A sleepy world of dreams.

SWINBURNE.

THIS is the time when the longed for cold weather seems illogically further off than ever, giving no hint of its approach, even at earliest dawn. Yet there is a certain charm in these hot, splendid mid-September mornings, when the dew lies thick on the grass until the sun has risen high, and the rarefied air is so clear it is possible to see for miles across the flat, humid country.

On such a morning lately the glorious sunshine and the slow breeze that moved the tops of the trees tempted me out of my garden into the wilderness beyond. I drove out through a part of the country that the floods had barely touched, through fields of Indian corn, now ripe for cutting, of vividly green rice plants that shuddered in the gentle wind, of freshly cut indigo, with its pungent, cleanly odour. The dome of the sky was deepest turquoise, thickly flecked with snowy clouds that turned to burnished silver where the sun touched them. Along the horizon the sky faded to a delicate blue, and against it masses of pale smoke-grey clouds were piled. Such a gorgeous morning of blue and silver and gold, full of life and movement and promise; far more tolerable despite the burning sunshine than the dull days we had last month, with sad evenings that close in with rain.

I avoided, as far as possible, the villages. If you are sensitive to the sufferings of dumb creatures a drive through an Indian village means a good deal of mental pain. The sight of cattle tied in the sun—the tightly hobbled donkeys, the miserable dogs, and over-burdened ponies, stab at the heart. None of it is wanton cruelty; just the callousness that is born of lack of imagination. We meet it in other forms amongst our own class, and frequently we christen it "tactlessness."

Better almost to be born without a limb than without imagination. The child who is devoid of it has a dull life indeed, for no bought plaything can rival the joyous devices of the imagination. I knew once a four-year-old boy who, on being assured that there was no such thing as a "booyah" (his term for ghost), gravely assured me that he had that very evening run one to earth in the verandah, and that having been dislodged from behind a flower pot with his father's fishing rod it was now sitting in the garden well, playing "Will ye no' come back again?" on the bagpipe! I don't suppose imagination could go much further than that, and I gave up my conscientious attempt to disabuse his mind of the supernatural.

Of course, I did not have the sun headache that was predicted for me. I have noticed that no one ever does have headaches when they are doing the things they want to do. Which fact seems to suggest that to be perfectly happy and healthy we should never do those things we do not want to do. I have frequently thought that if I could choose my disposition I would choose to be a thoroughly selfish person. But, of course, unless you are born with it, selfishness is one of the vices that it is practically impossible to cultivate with any degree of success, however much you may envy the triumphal progress through life of those persons who not only know what they want, but see that someone else obtains it for them.

The dogs did not approve of my desertion of

SO SHORT A TIME AGO . . . HE WAS A JOYOUS PUPPY.

them. They sat in an incongruous row on the steps and endeavoured to look contemptuous; the dachshund almost decided to beg, but thought better of it. They were not really uneasy, because they knew that no packing had taken place. Packing they view with a distaste that develops into resignation as the boxes are carried forth. The filling of a suit case can be endured with fortitude; at the worst it means a night's absence, and there is always the joyous possibility of an evening return; but prolonged operations terminating in the despatch of a large travelling trunk reduces the pack to a condition of melancholy and reproach that it is difficult to witness with indifference. In the case of the dachshund, this frame of mind is tempered with hope, she being the only one of the three who is ever taken away from home. To this end her fixed determination is not to be overlooked, and her devious methods are apt to be extremely disconcerting to the packer; should she be successful, the first mile of the outward journey is accomplished to the accompaniment of squeals of joy from the usually placid and sausage-like person, varied by excited efforts to fall over the splashboard on to the horse's tail.

So far as my garden is concerned I am, at present, engaged in the prosaic occupation of endeavouring to raise cabbage and tomato plants, Cape gooseberries, celery, and artichokes; vegetable marrows, too, which are not usually a success in Behar, did very well in my garden last year. They have to be cut before they become absolutely ripe, as a small insect then attacks them. I don't know if strawberry plants are ever raised from seed in these parts. I sowed some, but it has not yet come up, and I doubt if it ever will.

THE LITTLE WISE WHITE FIGURE.

My chief present joy is in my shrubberies, which are looking glorious just now, and should look far better next month, and in November when the yellow chakoor will blossom against the bronze background I have prepared for it, and the canna will be at its best. For I have never found that canna does best in the rains. The early cold weather seems to see it in fullest perfection, and, indeed, I think all Indian gardens are at their loveliest then. By January, when the English flowers are in bloom, the grass has lost the green memory of the monsoon days and dust is beginning to gather on the shrubs.

One great regret as the rains come to an end is that I did not plant more trees. I worked hard, got many plants into my garden, battled with floods and the battalion of insect pests that floods leave, but I did not—does any one ever?—do all that I meant to, or even one half. There were some delightful silver-barked trees—Australian, I think—of which I meant to plant a little grove. They grow very quickly, very straight and tall, and have vivid lettuce-green foliage. They also have the convenient property of attracting to themselves all the saltpetre in the adjacent soil and thriving on it, which certainly makes them trees to be cultivated in Behar.

I found an old book on 'Domestic Floriculture' the other day—a volume that, in my extreme youth, used to fill me with wonder and delight, so deep was my admiration of the atrocities therein depicted. The fern cases and aquariums, the stiff corsage bouquets, the dyed grasses, the gilded foliage, the chimney piece vases, are all illustrated at length, there is even an unpleasant looking bower of ivy, designed for the embellishment of a drawing room, the ivy having its root in a narrow china trough that encircles the bower. I wonder did any one admit these things to their drawing rooms, even in the sixties, when clothes and furniture alike were at their ugliest and most ornate. But, after all, beauty is largely a question of fashion and habit, and the future generation will, quite certainly, sneer at our taste for silver tables and art furniture, and the general futility of our designs.

All books about gardens appeal to me, but there is one that clings to my memory more persistently than any other, so that I can at any time recall long passages from its pages. It is a story of the early settlers in Kentucky—a story of courage, passion, and renunciation; and I do not think any one can read the book and not be touched by the tenderness and fragrance of it. And apart from the human interest the pictures of that green "land of promise" are entrancing. There a lonely woman made a garden.

"That year she sent back to Virginia for flower seeds and shrubs and plants—the old familiar ones that had grown on her father's lawn, in the garden, about the walls, along the water—the flags, the lilies, the Virginia creeper, lilacs, honeysuckles, roses—all of them."

"I think I know now the very day you will be coming back," she wrote from that garden to the man who never did come back; "I can hear your horse's feet rustling in the leaves of—I said—October, but I will say November this time."

He never came; but years afterwards, in the autumn of his life and the winter of hers, when the plants had become old bushes and the Virginia creeper had climbed to the top of the trees, he wrote to her . . . . "Ah, I did come back! I have forever been coming back! Many a time, even now, as soon as I have hurried through the joyous gateways of sleep I come back over the mountains to you as naturally as though there had been no years to separate and to age . . . . Ah, Jessica! Jessica! Jessica! . . . . and to this day the sight of peach blossoms in the spring . . . the rustle of autumn leaves under my feet . . ." That was all, for the woman who had kept her beauty for him, so that "to the last youth of her womanhood it burned like an autumn rose which some morning we may have found on the lawn under a dew that is turning to ice." Perhaps it was enough; the real women ask so little — just remembrance.

CHAPTER XV

OCTOBER.

Now gird thee well for courage,
My knight of twenty year,
Against the marching morrows
That fill the world with fear.
* * * * *
Yet fear them not! If haply
Thou be the kingly one
They'll set thee in their vanguard,
To lead them round the sun.

BLISS CARMEN.

THE English March, which traditionally blusters in like a lion and makes its exit with lamblike placidity, is nought in its variability to the Indian October, coming as it does with days of leaden skies, of steamy heat, or of scorching, pitiless sunshine, with scarcely a promise, even at dawn, of the gracious days that close the month. But now, in these last days, we wake to misty mornings, and the roses on the breakfast table are pearled with chilly dew; a cool breath comes across the smoke-wreathed fields in the evening, and it is good to remember that all the glorious days of November are close ahead of us. To say farewell to punkahs, and to welcome in the whitewashers, to sow the English flower seeds, and to watch the chrysanthemums bud—these are worth having waited for through long hot months. Better than all, though, I rank the practical certainty of "sleep in the night," one of the gifts of the gods to man, but, in India, how elusive!

These golden calm days with little soft breezes

ZINNIAS ARE COLD, PRECISE, SELF-CONFIDENT FLOWERS

out of the west make me think of English midsummer; of poppy fields, and sapphire seas, and red cliffs. In my very early days, before I migrated to the land of eucalyptus and wattle, I played one whole summer in that "Garden of Sleep" which provided the theme of Isidore de Lara's charming but now hackneyed song. I gathered poppies on those "green graves of dear women asleep"; and though my memories of those days are misty and distorted, as are all the memories of childhood, it seems then to have been always fine and always afternoon; very still, too, with only the low hum of bees and the hot pungent scent of poppies on the clear air.

It seems that we never had cold, wet summers in the days of our youth; we never acknowledge them, anyhow. Time is kind and eliminates our painful memories. We never forget the happy days of our lives, but the instinct of all human beings is to turn away from the memories that hurt, and so, uncherished, they die at last. I suppose that is how we manage to endure life.

I will never again sow English flower seeds before October. My dreams of unlimited carnation plants for bedding out are but dreams! The seed sown in August refused to germinate at all, and the September sowings did so only under protest, and then proceeded to damp off. True, the wall-flowers are alive and have in three months attained

THESE CALM GOLDEN DAYS.

the majestic height of two inches; so I think I may be excused for concluding that the results do not compensate for the anxiety and labour involved. I am now sowing sweet peas, nasturtium, candytuft, and the like, cosmos also. Of course my earliest sowings of this were a disappointment, though last month's sowings are now yielding me plenty of decoration for the house. I am especially delighted

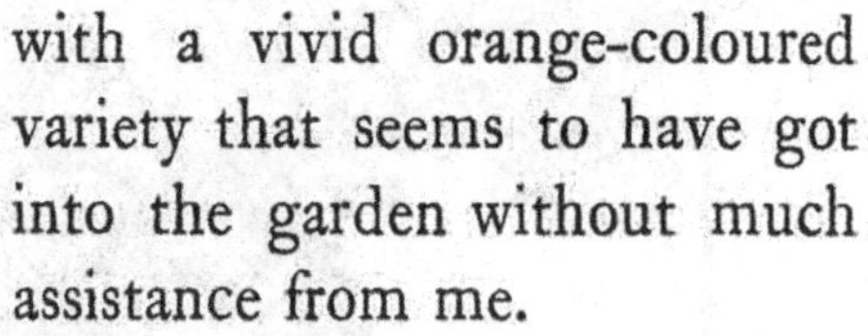

with a vivid orange-coloured variety that seems to have got into the garden without much assistance from me.

In Indian gardens we persistently neglect the humble marigold, the "gainda" beloved of the *mali.* Certainly the scent is unpleasant, and the flower, as seen in wayside gardens, not especially beautiful. But the large double variety, in every shade of yellow, from palest lemon to orange, are really beautiful, and last year I used these for filling up the spaces in a shrubbery in which, owing to the malignant influence of a teak tree, since removed, few other plants would flourish. Of course, the *mali's* tendency to arrange them on the dinner table had to be sternly suppressed; and indeed they are at all times more enjoyable at a distance.

The roses really are roses now, and not mere rosettes. I am waiting for the first really cold nights before opening the roots; but they will not be left open. Most Indian gardeners are especially determined on this point, but in a country where there is no hope of frost there is no object in exposing the roots at all. So mine shall be filled in when they have been well manured, and when any old wood or suspected stock has been cut away they will be watered every day. I have decided to leave a good many of them where they are, and not bring them into the long border in the front part of the garden. I am afraid of the trees, for roses should be well away from trees, and there is a large sissoo close to the house that looks as though it would become dangerous once it realised that good soil was to be found round the roots of the rose bushes. It really is a puzzle to know where to put roses in this garden, for they must have sun, and all the places where they would look best and are most needed are far too shady, the arbour for instance where the Sandwich Island creeper is now abloom in its glory, although a little oppressed by some mauve thunbergia, which is a rough, grasping creature, only forgiven by me because it happens to flower in my favourite colour. I acknowledge that I planted too many creepers through sheer impatience, and the desire to make a new garden look as cultivated as an old one. The rustic arches are all covered now, and the problem is how to cope with the too luxuriant growth, and where to put the creeping roses.

The dogs have, at present, a canine visitor, in the shape of a silken-haired amber-eyed Lhassa terrier, treated by them with politeness and reserve. The alien from far Thibet in no wise resents this attitude, being himself reserved to the verge of indifference; a gentle, sweet-tempered, intelligent creature who has the thoughtful appearance of one engrossed in deep affairs, and who trots at the head of the pack with a suggestion of "swagger" in the swing of his fluffy hindquarters. He has incidentally an owner; one of those restful persons who realise the important place that dogs and gardens take in the scheme of life as it has to be lived, and who ask no more of a hostess than a comfortable chair, a choice of books, and a little spontaneous conversation.

One could wish that all visitors should be contrived on the same plan, but alas! to be a successful visitor appears to demand certain exceptional qualities, and there are few people who could not write a heartfelt disquisition on the subject of "Visitors I have met."

* * * * *

I am always a little annoyed with people who openly wonder "what you can find to do with yourself all day" in the isolated life most of us lead in Behar. The implication is that you must be a dull person if so dull an existence contents you. But I honestly believe that dulness comes from within, and if you have the elements of it in you you will be dull in the most brilliant society just as surely as by yourself. Even boredom is relative, and you may take it as certain that if you are cast with people with whom you have no affinity they will be as bored with you as you are with them.

By affinity I do not mean anything sentimental. I believe in affinity in acquaintanceship just as much as in friendship or in love. It is probably more necessary than in the latter case, for when we prate of affinity in that sense we are apt to overlook the natural affinity—the mere unconscious physical attraction—that every young person has for another of the opposite sex.

This question of boredom is a large one, and to those who are at all given to self analysis it comes as an alarming fact, that no bore, of however virulent a type, ever recognises himself as such, and is quite likely to decry in others his own obvious characteristic. One can only conclude that it is this ignorance of our own most salient qualities that reconciles many of us to existence.

PANSIES.

CHAPTER XVI

NOVEMBER.

To travel hopefully is better than to arrive; and the true success is to labour.

R. L. STEVENSON.

In these November days, when I have at last leisure to rest and look about my garden to see the results of the past months of labour, I realise the truth of the above sentiment. Now that my work is practically at an end, and my garden glows in masses of colour, of bronze and scarlet, gold and blue, with poinsettia and chrysanthemums and morning glory, I almost regret those strenuous days of work. Yet my garden is nearly all I meant it to be; if my early sowings of pansies and carnations failed the later sowings are coming on now, and most of my last year's carnations survived the rains. I have pots of geraniums raised from seed, some violets, and heliotrope, and more wallflowers than I can conveniently plant out. The dahlias, too, look promising, though they did not really begin to show much interest in living until the nights grew cold.

I have been busy transplanting my roses, and after a good deal of mental indecision I have removed all the passion flower, convolvuli, and bridal creeper from the arches that span the path to the west of the house, and have planted all my creeping roses there instead, arranging stands, or a species of fence between the arches, so that the roses may continue their career between them, and so, in time, make the whole walk a mass of roses. It does not take very long to achieve an effect of this kind in India, this country of swift growth and as swift decay.

There are six roses on each side of the walk, amongst them a real Crimson Rambler which I am not sure of ever having seen in the plains. There is a climbing variety of both La France and William Allen Richardson, and an "anemone" rose that I am rather specially anxious to see, for I have been told that it flowered gloriously in Lucknow. Souvenir de Madame J. Metral is another climbing red rose that I have seen in flower once or twice, and for pink roses I have Dorothy Perkins, Reine Marie Henriette, and a daughter of hers, Madame

Driout, who is said to surpass the mother rose in beauty, but whom I have not yet seen in flower.

I wonder, at times, why we in India have contented ourselves for so many years past with Maréchal Niel, Gloire de Dijon, and Cloth of Gold as our only climbing roses. They are beautiful things, and will always retain their supremacy; but there is infinite delight in discovering new varieties and new effects, and every dewy morning when I go out to inspect the half-opened buds I do so with a delightful sense of expectancy that repays all work and anxiety. The other roses I have left to the east of the house, where they get plenty of sun and are away from trees. Amongst the really beautiful roses that the cold weather has brought into perfect form is a deep golden yellow rose called appropriately Perle des Jaunes. Soleil d'Or is another lovely yellow rose, very sweet scented. Other good yellow roses are Mrs. Myles Kennedy, Madame Philippe Rivoire, Mrs. Peter Blair, and Madame Constant Soupert. Amongst the pink roses I admire most are the Archduchesse Maria Immaculata; Catherine Mermet, who is, of course, extremely well known; H. Armytage Moore, a new rose, flowering in deep carnation pink. Frau Lilla Rautenstrauch is a delicate pink rose with a deeper pink heart. "Etoile de France" is the best red rose I know amongst the hybrid teas, and others are Madame Lambard, Gruss an Teplitz, a glorious red, sweet-scented rose, and Richmond. W. E. Lippiatt is another crimson rose; but I must confess that I think the French roses run far more to delicate yellows and pinks than to the glorious crimson shades that I love so much. I have all these varieties, and many more, flourishing in my garden at present, and by January I hope to have bunches and bunches of flowers to bring in for the house—but I have determined that I will not look forward to January. The cold weather is so short that it is unwise to anticipate a single day of it; we may lighten and support the hot weather by "looking forward," but in the cold weather we must live absolutely in the present. It is difficult, indeed, to believe that this respite is only for a season, for it seems as though these cool, sweet days must last for ever, and that no hot winds can ever again wilt the garden and parch the fields.

The chrysanthemums are not yet in full flower, but in a week the garden should be all aglow with them. I am afraid real gardeners would not think much of my chrysanthemums, for I have not the heart to nip off very many of the buds, and they are allowed to flower at their own sweet will. I do not like to seem unkind to flowers, any more than

I like to suppress the laughter of children or the gambolling of puppies, which are to me both precious and delightful things. I have coleus in quantities to mix with my chrysanthemums; the cold nights have touched the leaves to crimson, and, as I have planted it all about the house, the effect against the pastel yellow of the pillars is very good. The uniformity of the arrangement is somewhat disturbed by the dachshund's habit of projecting her clumsy self from the verandah in passionate quest of squirrels. This descent of a heavy body, which almost invariably occurs on the top of my cherished plants, is responsible for several breaches in the line, and causes the man of the house to drag up an ancient platitude to the effect that dogs and a neat garden are distinct incompatibilities.

IN PASSIONATE QUEST OF SQUIRRELS.

I have been lately tending upon a sick dog, and I know of nothing better calculated to quicken the sympathies and to eliminate all that is hard and distrustful in the mind. A dog puts himself so unreservedly into your hands when he is suffering; he gives you to understand that all responsibility lies with you; and he never, for a moment, doubts that you are doing your best for him. Your mere presence comforts him; now you may taste of omnipotence, for to him you are all powerful. Indeed it seems to me that it must be a sad, cold nature to which dogs make no appeal—that can ignore the love and fidelity, the sympathy and the trust that the dog is waiting to give. Certainly to those who love and have understanding of dogs, life can never seem utterly hard and dreary.

THIRD YEAR IN THE GARDEN

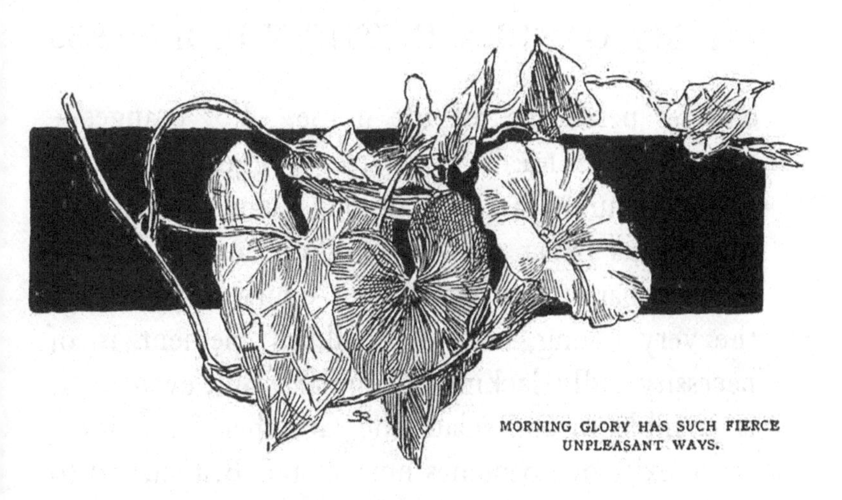
MORNING GLORY HAS SUCH FIERCE UNPLEASANT WAYS.

CHAPTER XVII

JANUARY.

Then you will find—it's your own affair—
But. . . You've given your heart to a dog to tear!
R. KIPLING.

IT is at Christmastide that the heart of the exile is supposed to turn with deeper longing to his native land. But, in Behar, it is possible to have so fair an imitation of an English "green Christmas" that I am inclined to think that it is at this season that most of us feel our exile least. The chill of the early mornings is akin to frost; there are big log fires burning in the evenings, and often in the daytime also, for this year cold winds blow across the wilderness. There are late, lazy mornings in bed; there are masses of chrysanthemums in the drawing-room, and the peculiarly festive scent of oranges pervading all the house. For oranges—eaten between meals, of course—suggest Christmas just as surely as do almonds and raisins, or mysterious paper parcels, or chestnuts roasted on the fire bars. Christmas is of course a festival of the very young, and the childish element is of necessity sadly lacking in India. But, even so, it is possible to appreciate the fact that for a short space existence becomes normal and British, and to find in the wreaths of poinsettia leaves and marigolds with which the *mali* seeks to decorate the verandah some suggestion of holly and of laurel.

My drawing room holds vases and cups of wallflowers now, and I am trying to behave as though I had always grown wallflowers, and knew just how they should be treated in the plains. In truth, no one of my neighbours is more surprised at this profusion than I myself. I took trouble with the seedlings, it is true, but so did I with carnations, and pansies, and many other annuals, and the results are not such as to arouse any enthusiasm. The scent of the velvety brown-gold blossoms that we value so little at home is a breath of England out here, and it is just because they are so common that they appeal so widely, for every one, even the least learned in floriculture, has known wallflowers, has seen them grow in cottage gardens, spring almost wild in kitchen gardens, along hedges, by the wayside. Geraniums, too, I have raised from seed. The plants are looking very well and healthy, but will not flower till the weather grows warmer, and I am now putting out my last sowings of carnations, which, also, should flower with the first warm days. The plants I kept through the rains, although healthy and well, have not shown the remotest sign of budding yet, and I can only surmise that Vienna carnations do not like this climate, and that the marguerite variety which I have hitherto cultivated with success are the most suitable for this part of the world.

François Crousse has not disappointed me and is at this moment covered with bright cerise blossoms, borne in trusses, very brilliant and perfect. In colour, and indeed in form, he closely resembles the Richmond rose. Reine Marie Henriette is also in bud, but I am not especially interested in her. I had her in my old garden. She is beautiful, but almost scentless and flowers grudgingly; a veritable termagant amongst roses. William Wood flowered for the first time a few days ago; he is not in the front walk, but in the rose garden proper, and is the darkest rose there is, I think, with deep purple shadows in his heart, darker even than the Black Prince.

As even the prettiest of women has days on which she looks almost plain, so gardens take on their seasons of something approaching ugliness. My garden looks, at present, its worst, and has almost the air of being conscious of it. The falling leaves of the sissoo and toon trees give a forlorn and autumnal look to the landscape, the grass is already tinged with brown, and now that the chakoor bushes, which last month were in the height of their vivid bloom, have dropped the last pastel yellow petal, there are but the poinsettias left to brighten the shrubberies, and to give the colour that I like so well. There is a bed of eschscholtzia at one side of the drive, that is in flower, or would be, if we did not have such sunless days; as it is the flowers keep their golden petals obstinately furled. The waves of damp white fog that drift across the garden do their best for it in giving an air of mystery to the scene. After midday the mist lifts and the sun comes out, and all the world seems still and warm, but with dusk the white bank rolls up again, and suggests early frost in the Highlands, the frost that ripens the blackberries, and touches the leaves of the wild cherry tree to scarlet.

This damp, soft weather that I rejoice in does not suit my little old terrier as it does me; it gives him rheumatism, as he signifies by holding up a piteous paw for my interested inspection. It is only with me that he shakes hands now, for strangers cannot be trusted to know and to consider the susceptible joints. As I watch him I realise sadly how completely I have "given my heart to a dog to tear." It is a pitiable thing to have done, but ah! how many have done it before me, and how many will do it yet! Those humans who have escaped the pain of losing a dog friend are scarcely to be envied, for they have also missed something rather precious in the way of love and sympathy.

It seems so short a time ago that my terrier was a joyous puppy, and now he has given in to the advances of age; he likes the comfort of the fender stool, and seeks no adventures when we walk. But he has had a good life; a dashing joyous life, full of interest and movement; a life of which he has lived every moment. What desperate fights with pariah dogs he has had; what zestful hunts; what gory kills; what glorious wounds and scars are his; what a life all round! The dachshund watching him, as she often does, in what appears to be respectful admiration, can never know such moments as he has lived through. Her life has been, and will always be, a sheltered one; and, in contrast to his slight body and hardened muscles, she already in early life presents the appearance of having been cut out in black satin, and very carefully stuffed.

* * * * *

Outside my garden there stretches a waving sea of gold that the prosaic mind recognises as mustard. When the west wind sighs across, it ripples and dances, and in the morning before the soft blue mist, that is characteristic of the cold weather here, has gone from the horizon, the sight is a glimpse into fairyland. Beyond the mustard there is a field of dull green tobacco plants, acres of them, it seems, for this is the land of tobacco. Further south they grow chillies as winter crop,

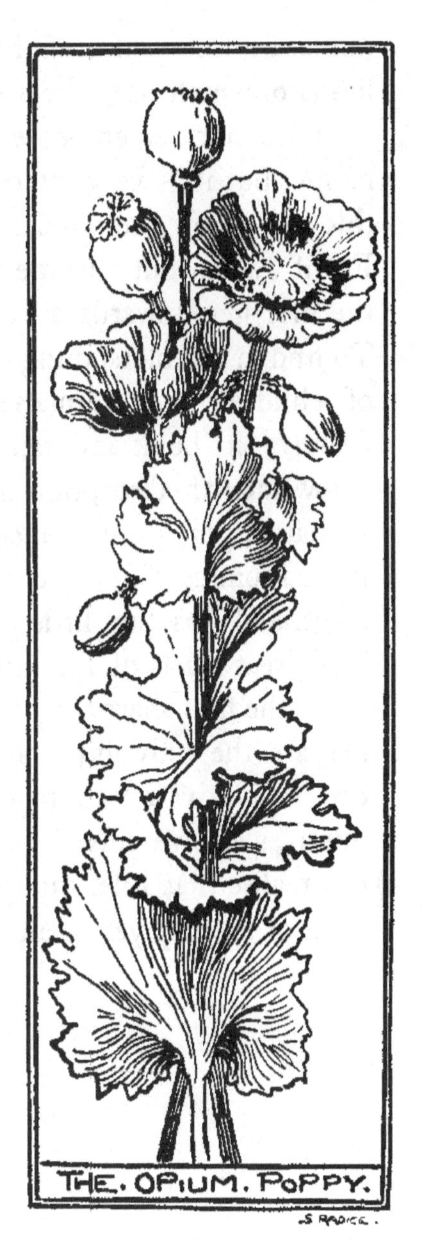

THE. OPIUM. POPPY.

and in February the roofs of the village huts turn to orange-red with their covering of drying seed-pods.

We do not cultivate opium in this part of the district, and at first I missed the fields of snowy poppies that, for a short time, relieved the monotony of the landscape. Tall, imperious things are opium poppies, but their life is short; for at the height of their beauty their petals are torn from them, their sides are scored, and in one day all their pride is over, and the field that fluttered like a swarm of white butterflies is but a miniature forest of straight, bare stalks, from the heads of which ooze the sleep-giving drug.

I do not remember who said that "life is not happy, but it is very interesting," but it seems that this is not, on the whole, a bad motto to accept. For if we can but become interested we have taken the first step towards a certain kind of happiness. To find in the blossoming of a garden; the gambols of children and animals; in the confidence of friends; in book and talk, and the society of our fellow men an absorption that arises from sympathy —that is, going out to meet happiness on the way. For happiness is almost entirely a matter of temperament, and has but little to do with environment. The great griefs of life threaten to overwhelm, but when the first sharpness is over we find the trivialities are the only important things, and that if we can but cultivate "courage and gaiety and the quiet mind" we have solved some of the problem, and can realise that Life, though but a sad business for most of us, may yet be an extremely interesting one.

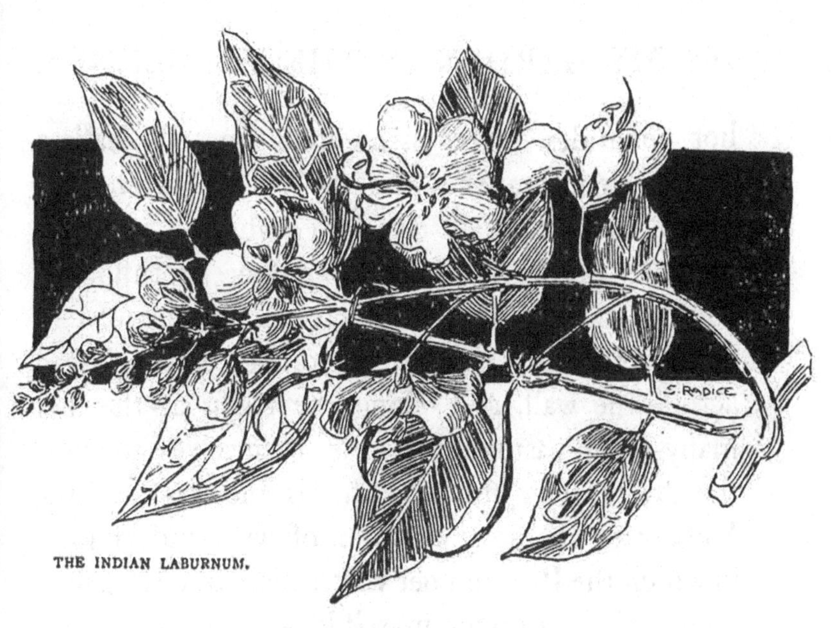

THE INDIAN LABURNUM.

CHAPTER XVIII

FEBRUARY.

My garden is a lovesome thing—God wot!
Rose plot,
Fringed pool,
Ferned grot—
The veriest school
Of peace: and yet the fool
Contends that God is not.
Not God! In gardens! When the sun is cool?
Nay, but I have a sign—
'Tis very sure God walks in mine.

T. E. Brown.

February in Behar is the month of flowers. So long as the west winds are lulled it is an ideal month, with a sharp chill in the morning air, and hot noontides. The sissoo trees are still leafless, but in their branches the *koil** is awake and the doves coo softly all day long, while the garden beds are gay with a tangle of poppies, cornflowers, cosmos and the like. The creeper on the western end of the house is like an orange flame spread across the wall, and along the trellis at the foot many-hued nasturtiums raise themselves to meet the tawny glory above. This is the season of the Rubaiyat—the song of roses, of wine and of love, in which the Persian poet voices his attractive pagan philosophy. I do not myself love that sweet shallow singer, but in his jewelled verse he has caught the true spirit of the Oriental spring, and I like to wander in his garden by the river brink in "this first summer month that brings the rose." And truly it does bring the rose, for now is the time when it seems, at last, that all the endeavour, the time and the care spent on the rose garden has not been in vain.

All those roses I planted in the front walk have now flowered with the exception of Dorothy Perkins and the Crimson Rambler, who on one of the rustic arches face one another in obstinate quiescence. Madame J. Metral, who closely resembles Reine Marie Henriette, is covered in

* The Indian cuckoo.

heavy crimson blossoms, and there is a red La France (of 1887) that is beautiful, though not taking the scent of its pink relation. I am, I think, just a trifle disappointed in Reve d'Or, which has just flowered on the nearest arch, a creamy yellow rose with a smear of pink on the reverse of its petals. It is not such a "dream of gold" as is the Duchesse d'Auerstadt, who flowered for the first time yesterday, and is beautiful beyond compare; in form and leaf reminiscent of the old Maréchal Niel, but in colour, gold—old gold, with no hint of copper in the shadows of her lovely heart. I have seldom seen a rose I like better, and I am glad to know that the plant is covered with buds, so there is nothing grudging about her. In admiring her I forget to thoroughly appreciate an effect that has appeared on one of the trellises that connect the arches — a veritable Burne-Jones background formed by the Sinica rose, an anemone rose that flowers close to the stem, and is now covered with wide pink single blooms like monster dog-roses, and having a delicate honey scent. It is worth cultivating for this single season of fleeting beauty, but it is one of those parsimonious roses that flower but once in the year, and such have no real place in my affections; they remind me of the women who are gracious only in society.

But spring in my garden is not all Oriental. Outside my dining room windows there is a south corner that is entirely English in design. Here a screen of mighty hollyhocks shuts out the vista of kitchen garden, and the sun strikes hot upon wallflower and mignonette, planted in orthodox fashion against the wall, and on a row of sweet peas that are just coming into flower. So here is a mingling of fragrance to be wafted in at the windows while we sit at meals. They are real windows, not the window-doors to which we are accustomed in India; four of them, looking on one side to the rose garden, and on the other to this fragrant nook. The room is quaint and low-ceiled, gay with blue china against the dull red walls, and comfortable only so long as this cool season lasts. For, in India, comfort does not go hand in hand with the picturesque, and the border where the wallflowers now glow will be arid and empty a month or so hence, whilst the burning winds that blow across the parched wilderness beat hard on the unprotected walls.

Between the wallflowers and the sweet peas life is really rather a strenuous business at present; for if they are to continue to make the garden gay for any length of time, they must be plucked and plucked, and here is a labour, even though a labour of love; for I do not care to entrust all this to the devastating hand of the individual who represents himself to the rest of the household as the *mali*. I do not so designate him myself, his own unsuitably picturesque name of "Golab Chand" or the "rose moon" helping me to evade the difficulty that was

THE TENDENCY TO ARRANGE THEM ON THE DINNER TABLE.

thrust upon me by the apparent impossibility of finding a *mali* who would carry out orders, and make no disastrous efforts to think for himself. Since receiving a coat, an infinitesimal rise of pay and permission to occupy the grass hut that yearly threatens to slip from the edge of the garden into

THE GRASS HUT THAT THREATENS TO FALL INTO THE RIVER.

the river, Golab Chand has naturally eschewed the weeding and watering that has been his daily occupation for some years. But he is zealous in seeing to it that less favoured individuals perform these tasks, and on the whole I prefer his cheerful unintelligence to the supercilious attitude of my late *mali*, who delighted to inform me that his last *sahib* or *mem-sahib* employed gardening methods that were absolutely foreign to my own. Of course they did! Only strictly unoriginal persons would garden on the same lines as any one else. That is why gardening books, although delightful reading, are never of very much use to anyone, and our experience provides the only really reliable information. I could never be really content in a garden developed according to some one else's ideas, any more than I could buy a piece of fancy work begun in a shop "with silk to finish" and make it into an adornment for my house. The garden and the needlework might both be beautiful, but it seems to me that gardens as well as houses must have individuality to be really attractive.

I do not like flower-shows, and I do not suppose that any real lover of flowers does like them. I appreciate the warm, hot-house-like fragrance of the air, that indescribable scent of damp earth and flowers, and fruit that seems to belong to such shows. But I am always intensely sorry for the flowers, which are reft of their suitable surroundings, arranged in unbecoming rows, and subjected to criticism that seems more than cruel. Judges with uncompromising eyes, and merciless note-books comment harshly on their form, their texture, or possibly their lack of heart, while the roses, which but that morning opened hopefully in the dew, now droop beneath the heat of the room, and the contempt of their fellow roses.

I would have a flower-show a real show of flowers: just that. They should be massed against a suitable background, and there would be no judges, no first prizes, no wicked human-like striving to outdo, no heartburnings, no jealousy, just a veritable display of all that is lovely in the garden. But, of course, such will never be until there arrives that millenium of peace for which we all pretend to hope, and with which we should all be intensely bored should it ever come. I have no doubt that most of us would enter into the feelings of the puppy of which Barry Pain tells us somewhere, who, after death, found himself in "a long quiet street," where he realised sadly that he no longer had any desire to chase cats, or steal, or do any of the bold, bad things he had done in life, because he was in the place where good puppies go to. In the end he decamps towards where snarling and howling proclaim a glorious fight amongst the bad puppies.

Speaking of bad puppies recalls Cecil Aldin's delightful 'Dog Day,' and so to the latest addition to our household, a rough-haired fox terrier, whose looks and expressions are very like those of the wicked hero of that charming book. She came amongst us but a week ago, and in an hour had determined her exact place in the family. She is gentle and unassuming, but I cannot help suspecting her of basing a certain superiority on the fact that she was born in England. For all that she is, being of a lively disposition, willing to frolic if she can but get some one to frolic with. The dachs, although youthful, is too "fat and scant of breath" to follow any game beyond the initial stages; my terrier, infirm in body and in temper, ignores the arch blandishments of the little lady, and the Shabby Widow, with her sedate shamble, is too obviously of the genus described by Stevenson as "respectable married people with umbrellas" to arouse even the faintest hope of a game. There remains but Sarah Jane, the white rabbit who spends her days in a wire run on the grass at the side of the house. The new-comer still hopes that some day Sarah of the opal eyes will arouse from her lethargy, and to this end she gyrates persistently before the wire-netting and archly waves a paw at intervals. Neither dogs nor humans have any terrors for Sarah Jane; she has lived with

SARAH JANE OF THE OPAL EYES.

us for over a year, and is thoroughly domesticated, even to the extent of preferring scraps of cakes and puddings to orthodox bunny food. She is sleek and fat and self-respecting, and the dogs treat her with consideration; so that a frolicsome puppy inviting her to go out and have a game means nothing to Sarah Jane.

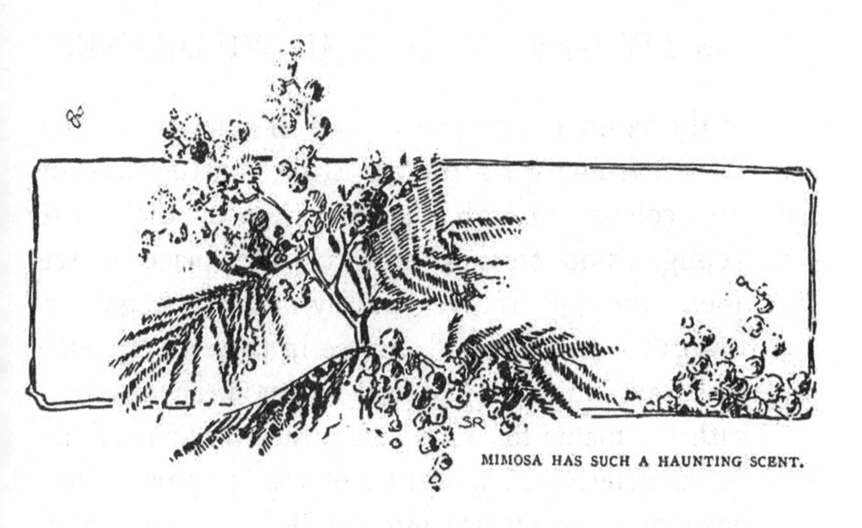

MIMOSA HAS SUCH A HAUNTING SCENT.

CHAPTER XIX

APRIL.

Oh, the birds, the tree, the golden
And white blossoms sleek with rain!
Oh, my garden rich with pansies!
Oh, my childhood's bright romances!
. . . . . I see them stir again.

E. B. BROWNING.

SPRING lingers late in my garden this year; even now the poppies and hollyhocks defy the heat and drought, and the lower portion of the garden, by the river, has broken out into a tangle of mauve and yellow, in the shape of cosmos, and a tall yellow daisylike thing that I sowed broadcast in the rough plots down there. This part of the compound recovers from flood effects too late to secure much of my attention, so the effect is more in the nature of accident than of design; but in the early morning when the sun strikes on the mass of soft colour through the branches of the vivid young sissoo trees with which the place is set about, the sight is a very lovely one. The garden is worth getting up early to see in these days; and that says much, for I have never been obsessed with the mania for early rising that is reckoned to be so determined a feature of life in India. Just now the middays are often dulled with dust and heat and glare, but the mornings are incomparably fresh, and give such an impression of cleanliness and health that I feel it my duty to absorb as much of them as possible. But I like best to lie late in bed, with the scent of belated wallflowers and a babel of bird cries coming in at the open window, and to watch the broad plantain leaves swaying across the square of turquoise sky that bounds my early morning vision.

It is true that the lawns are brown, and the shrubberies dry, that outside my garden the wilderness stretches arid and grey, and that the great leaves of the teak trees hurl themselves about the compound, yet still my garden yields sweet store. The roses, although smaller, are plentiful still, and there are sweet peas and carnations, petunia, hollyhock and phlox in plenty. It is almost difficult to know what to do with the profusion that the *mali* daily brings in; every bowl and jar in the drawing room is filled, and even if I had more vases available there is really nowhere that I could place them. Just at this time, two years ago, when I first came to this place, I was seriously contemplating decorating my room with the young branches of the sissoo trees, so serious was the

dearth of flowers and so much do I dislike to be without some substitute for them in my room. I would be more than human did I not look about my demesne and draw comparisons between then and now; but I know all the same that my garden is not in the least as I then designed it should be. Gardens have an astonishing way of growing away from one and laying themselves out according to their own notions.

For some freakish reason that I cannot quite grasp, the chakoor shrubs are in full flower again. I was just going to have them pruned when they burst forth in this astonishing fashion. I find that mine is the only garden in these parts that possesses these exceedingly hardy shrubs, and I wish that I knew a more classical name for them, for they are so rapid in their growth, so good for filling up odd corners, and so ornamental at all times, that I want to recommend them to everyone who is making a garden. *Tecoma stans*, which is also in flower, strikes another note of yellow in the shrubberies, and the quisqualis that clothes the west wall of the house, and is now clambering madly on the roof, is in riotous bloom, and in the evening floods the air with its almost too penetrating sweetness.

My garden has positively rioted with sweet peas this year, and now that the ordinary pink and purple variety are nearly over there have come into flower those I reared from English seed. There is a vase of them before me as I write, deep crimson and white, pale mauve, cerise, pastel yellow, very large and sweet scented for such late blooming flowers.

The present fashion for sweet peas will never touch us very seriously in India—in the plains, at all events. The imported seed flowers too late in Behar to give us any great profusion of blossoms; and, try as we may, the second year will see us with little but the pink and purple pea of the cottage garden, and all our carefully saved and isolated seed will count for naught.

* * * * *

I have been wandering in other people's gardens; an occupation alternately provocative of admiration, delight and envy—I mean, when they are real gardens, as these were, although poles apart in design and intention. The one, with orderly parterres, had only escaped ribbon borders by the exigencies of the climate. The carefully bricked beds were filled with the flowers they were meant to show, a mass of purple petunias in one, pink phlox in another, glowing portulaca in a star-shaped bed. The lawns were apparently of velvet, and there was a wondrous arrangement of rose-arches copied from an old English garden. It only wanted yew and laurel clipped into urns and obelisks and peacocks to complete the effect.

The other, one of the most notable gardens of Behar, was on a grander scale, with well thought out shrubberies and vistas, and a natural fernery shaded in with palms, a seemingly endless maze in which we went amidst banks and masses of maidenhair and rarer ferns, of velvet-leafed begonias and silver caladiums and glowing coleus, while overhead the tall palms met, and all was dim and mysterious. There were rare plants there, but there were also common ones, and all had their sphere and found their uses. A bed of starry yellow stonecrop brightened a dull corner, common red spinach was clipped to form a border where its colour was most of value; nothing was despised—even weeds were made the most of. Years of care and thought had, of course, gone to the making of that garden, but it is possible that the maker of it was happier in those years of labour than now in the quiet possession of it.

* * * * *

At present life in a bungalow is being made wearisome by some beings called "gharamis" who clamber about the roof like monkeys. They are supposed to be supplying some deficiencies in the tiling of the bungalow, but their chief occupations appear to consist in hurling tiles and bundles of grass from above, thereby making life both dangerous and difficult for the dogs and the servants, and in discussing their affairs in loud drawling tones that penetrate everywhere. For some inscrutable reason no one ever says *chup!* * to a "gharami," and their activity always becomes peculiarly marked between the hours of one and three when the occupants of the bungalow are supposed to be enjoying a siesta.

Even if your house be a pucca one you cannot altogether escape the ministrations of the "gharami" at this season; for he will then come and drag tatties and mats about with an ugly, swishing sound, on the pretext that he is putting them there to make the house cooler.

* * * * *

There came a storm in the night—first a warm, rain-scented wind, and then a quick sharp tempest of thunder, lightning and lashing rain that lasted but a short half hour. Then it grew still again, and the full moon came from behind the flying clouds and touched the wet leaves and branches to silver. This morning a cool wind is blowing across the damp garden, and the flowers have

* Be quiet.

obviously taken on a new lease of life, while the fell hot weather which but yesterday seemed at our doors has receded into the dim future. A scent of petunias and carnations comes in on the wind, the

THE COTTON TREES ARE IN BLOSSOM.

young sissoo and peepul trees have had their "gay, green gown" washed clean of dust, and it does indeed seem that "spring rides through the wood."

Even the vegetable garden, whither I repaired in search of green food for the lymphatic Sarah

Jane and my illnatured green parrot, had taken on a new, clean aspect. Bees were buzzing about the blossoms of the lichi trees, the tomatoes had shining newly washed cheeks, and some wind-sown scarlet poppies were nodding and dancing in a bed of what the *mali* calls "buckla beans." Outside the hedge that bounds this territory there is a dried-up tank, on the bleak banks of which some twisted babul and kuchnar, *Bauhina variegata*, trees have their foothold. Yesterday I scarcely noticed them, but to-day the pinkish mauve blossoms closely massed on leafless branches stand out gloriously against the darker foliage of the babul and provide a blot of colour in the wide, grey landscape beyond.

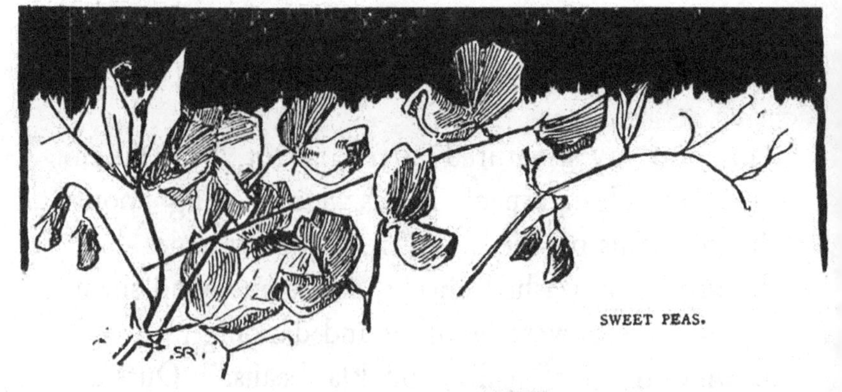

SWEET PEAS.

CHAPTER XX

THE LAST JUNE.

How sunshine beats on London streets,
I shut my eyes and know it:
An afternoon of early June,
The golden town below it;
The soft new airs in pleasant squares
The Park's sweet carpet thickened,
And hearts all out for holiday
Sun-gilded in their brave array.

N. B. TURNER.

IT is a far cry from the Wilderness to the little London garden in which I sit in the only available scrap of shade, and look up into a sky as blue and as hot as that of India or of Italy. In my Indian garden the sissoos have turned their darkest green, the teak trees are in blossom, and the river that bounds and frequently invaded my territory, is moving swiftly, swelled by the deluge of rain that brought in the monsoon. When I shut my eyes I am aware of it all; the sweep and hiss of the rain; the creak of the peepul branches as they toss in the gale that precedes the deluge; the song of the frogs in the pool beyond the hedge.

It is very quiet here; there is an air of peace about these little garden plots that does not exist in the big garden common to the square. A light breeze is rustling the leaves of the lime and elm trees that close in this scrap of green in London's heart, and causing the spikes of blossom on the Minnehaha rose that clothes the further wall to nod and dance widely. Minnehaha is a

THE RIVER BY MY GARDEN.

first cousin of Dorothy Perkins, the obstinate young person who refused to blossom in my garden in the East.

The lawn before me is the size of a pocket handkerchief, but there is a cool looking grotto of ferns in one corner, and against the grey walls grows a perfect blaze of geraniums, of calceolaria, of Canterbury bells and sweet peas, regal purple clematis hangs against the house, and there is a tub of hydrangea beside me. Every inch of space is made the most of. This garden holds memories. In the wall opposite to me, half hidden by flowers, is a plate "In memory of Jock, for many years the friend and companion of Mercy." It is good to know well that canine friend was loved. I wonder who Mercy was, and whether she is now grown up, and sometimes "thinks long" on that little faithful friend of hers.

There is a masonry slab now in my garden, and my little terrier sleeps well under the lichi trees in the corner. There is only one word there to his memory—"faithful"—for no dearer tribute can be given to him, and I am glad to know that his little, weary, white body is at rest.

"A glorious summer" say the Londoners: weeks of sunshine; skies almost brazen in their intensity; a wealth of flowers and fruit. Flowers are everywhere; masses of carnations and sweet peas such as I can never hope to raise selling "dirt cheap" at the street corners; flower shops with their more exotic wares; flowers in the parks; flowers in the milliners' windows. Here they frequently take on a form unknown to nature; purple roses, black or blue roses, wreathe the large hats worn with the dainty muslin gowns that throng the streets and the parks. Here is found fruit, too; emerald green cherries, royal blue plums, and a replica, also, of the real thing piled on hats as well as on barrows and in the fruiterers' windows. All is colour, a feast of colour in dingy old London, who is now, at last, divesting herself of her Coronation robe. And the girls in the parks look like

flowers, too. Englishwomen have at last lost their dread of bright colour, and are clothing themselves suitably on these glowing days. A vivid shade of cerise is this year the favourite; it gleams in the soft shade of the trees, and flashes by you in the street. The London girl, with her skimpy gown, coloured shoes and stockings, bare neck and large hat, has a fascination that the girl of the provinces frequently lacks. Often flimsy and tawdry as to garb, jaded in mien, she has yet the air that interests. She attracts because she desires to do so, which desire is, after all, the foundation of many women's fascination. Neither does she shrink from what is incongruous or unnatural, as her country cousin may, and I have seen a pretty, smiling vulgar face retain its charm under what looked like a straw flower pot decorated above the ears with bunches of radishes.

* * * * *

I have to my joy made the interesting acquaintance of several dachshunds, for the most part taking the air, on a lead, in the sad manner to which town dogs are reduced. The natural melancholy of the breed is increased under these circumstances, until I think it safe to assert that one ot the saddest of sights is a dachshund on the chain. One heavily built tan coloured person who resides in the same square has apparently come to regard my society as the chief alleviation to life; to the annoyance of my relatives, who regard as deplorable my habit of scraping canine acquaintances in the street. I believe they have visions of my arrest at some future date on a charge of dog-stealing.

To be a regular dweller in England seems to be a nerve-wracking business; he is always afraid of something. In dear, beautiful, safe England, away from the sudden tragedies that startle us in the East, he yet cultivates fear of many things—draughts, and the swoop of aeroplanes, sunstroke, and rabies, and the German invasion. To listen to him you would imagine dangers on every side. He demands standard bread, and believes he is being insufficiently nourished if he does not get it, and he thinks much about microbes, and such matters as we treat as a commonplace in India. Life in the East had some compensations after all.

Green and beautiful as London now is, my heart, at this season, turns to a garden I know of in the North, a garden set about a solid unpretentious house to which time has given both dignity and beauty. There grow strawberries such as I have never tasted elsewhere, and the wall fruit ripens lusciously and very slowly, because of the breezes that blow in across the most beautiful firth in the world.

"C'est beau!" Mary of Scots is said to have exclaimed when she looked across to the low purple hills and dark vales of the Black Isle; and Beauly it is now, that little village in the bend of the firth. And there I soon shall be, in that walled garden, where fruit and flowers and vegetables grow sociably together, and life can be, for the visitor, one long-drawn afternoon. I love London, but there is a charm in Scotland, too, and I like to think that soon I shall, after so many years, see again—

> "The hue of Highland rivers,
> Careering full and cool,
> From sable on to golden,
> From rapid on to pool."

I can give up all the glow and glitter of the English summer for the sight of the peat shadows in the Highland rivers, and the light on the heather hills above Beauly.

LONDON:
PRINTED BY WILLIAM CLOWES AND SONS, LIMITED,
DUKE STREET, STAMFORD STREET, S.E., AND GREAT WINDMILL STREET, W.

Printed in the United States
By Bookmasters